NON-LOCAL GAUGE FIELD THEORY

BY

BOJAN TUNGUZ

B.S., Stanford University, 1997
M.S., Stanford University, 1999

DISSERTATION

Submitted in partial fulfillment of the requirements
for the degree of Doctor of Philosophy in Physics
in the Graduate College of the
University of Illinois at Urbana-Champaign, 2006

Urbana, Illinois

Abstract

This is a study of Non-Local Gauge Invariant Quantum Field Theory. Non-Local gauge invariance is an extension of the group of local gauge transformations that has been extensively used in construction of physical theories. We show how this larger group is rich enough to accommodate a new theory of gravity, and we propose a new model of unification of gravitational and electromagnetic interactions. This new model of gravity has the advantage over the general relativity of being a quantizable theory.

Mojoj majci Zlati.

Then the LORD said, "Go outside and stand on the mountain before the LORD; the LORD will be passing by." A strong and heavy wind was rending the mountains and crushing rocks before the LORD–but the LORD was not in the wind. After the wind there was an earthquake–but the LORD was not in the earthquake. After the earthquake there was fire–but the LORD was not in the fire. After the fire there was a tiny whispering sound.
(1 Kings 19:11-12, NAB)

Acknowledgements

This work has been supported through teaching and research assistanships, as well as through fellowships. I would like to thank the UIUC Physics Department, the University of Illinois and Professor Yoshitsugu Oono for their generosity. I thank all my fellow Physics graduate students for all the years of friedship and support, in particular Nick Romero, Hector Garcia Martin, John Veysey, Michael Lawler, Ken Esler, Dyutiman Das, Dom Ricci, Jordan Vincent, Zigurts Majumdar, Andrea Esler, Eun-Ah Kim, Michelle Nehas, Jack Sedlair, Kapil Rajaraman, Satwik Rajaram and Prasanth Sankar. There are many others in the UIUC community on whose friendship and support I have relied over the years, and I would like to thank the following: Echo Leaver, Cesar Luna-Chavez. I thank my Croatian friends at UIUC for providing a little piece of homeland for me: Ivan Feitl, Goran Skosples, Tatiana Kuzmič, Katherine Sredl, Marko Madunić. I thank my family for all their love: my brother Dalibor, my uncle Ljubo and his family, and all my cousins and relatives. I thank my great-uncle Kreĭmir for th encouragement he's given me to pursue science. I thank the committee members for providing me with valuable suggestions and guidance: Michael Stone, John Stack, and Mats Selen. I thank my adviser Yoshi Oono for all the years of support and guidance. Most of all, I thank Sharmin Spencer, for all the love and support she's given me.

Table of Contents

1 Introduction

1.1 General Outline

The twentieth century has been a very successful period for our understanding of fundamental physical processes. We have been able to reduce all of the observed interactions to just four fundamental forces. Three of these, electromagnetic, weak and strong force are the building blocks of the Standard Model and they are theoretically described by the local gauge field theories. The theory behind these forces is very well understood, and there is a plethora of experimental evidence in support of it. The fourth force, gravity, is described by the General Theory of Relativity. This is a very different theory from the other three, both conceptually and in terms of available evidence. Experimentally, we have been able to directly test this theory only in the linear regime, and conceptually we've never been able to fully reconcile it with quantum mechanics.

There is one more overarching problem with the above theories, and that is the problem of unification. Ideally, we would like to have just one theory that describes all of the above interactions from a single principle. In the case of the Standard Model, this is somewhat accomplished through the choice of a sufficiently large gauge group. However, this still leaves out gravity.

The purpose of this research project is to explore two concepts, nonlocality and gauge invariance, and to try to unite them in a physically meaningful and operationally useful way. In this introduction we will give a brief history of these two concepts in theoretical physics, and try to motivate their unification. In particular, we will try to show that General Relativity (GR) already incorporates both of these notions, albeit in a very special way. Thus understood, nonlocality already has meaningful and observable physical effects. In the rest of this thesis we will construct a particular non-local gauge field theory based on an arbitrary unitary differential transformations and show how it relates to General Relativity.

1.2 Non-Local Field Theories

Ever since it was realized that in classical electrodynamics self-energy of a point particle is formally infinite, there have been attempts to introduce non-locality into theoretical physics. However, only with the advent of Quantum Field Theory (QFT) did these considerations become one of the central concerns of the research in theoretical physics. This was due to the fact that development of perturbative QFT was stalled for many years by the existence of these infinities. Most of these concerns were put to rest with the introduction of the renormalization procedure in Quantum Electrodynamics (QED). In the subsequent decades this program of renormalization was successfully applied to all fundamental physical interactions except gravity. This lack of progress in our understanding of quantum gravity has kept interest in nonlocality alive. There have been many approaches over the years to this problem, and it would be impossible to do them all justice here. I will, however, try to describe a few of the more important ones and compare the approach in this paper to them.

1.3 Gauge Field Theories

The principle of gauge invariance has a long and successful history in theoretical physics. It is the central principle around which all dynamical physical theories are constructed. In particular, this principle is well developed for the interactions of the Standard Model. There one obtains the electromagnetic, weak and the strong interactions by demanding that the action remains invariant under the local $U(1) \times SU(2) \times SU(3)$ transformation of the physical fields. The local non-Abelian gauge field theories have been well understood for quite some time. The method of constructing these theories was first introduced by Yang and Mills [4], and subsequently discussed by Utiyama [5], Kibble [7], and others.

General Relativity (GR) [23, 29] is also a dynamical theory based on a gauge principle, but it is a gauge principle that is distinct from the one employed for the construction of the Standard Model. The gauge group of GR is the group of all diffeomorphisms of the space-time manifold. In the usual formulation of the GR this group acts on the local coordinate patches, and the gauge fields relate the changes in coordinates from one patch to another.

It is necessary to point out that the coordinate transformations (diffeomorphisms) are distinct from Lorentz and Poincaré transformations. This is a subtle point that is oftentimes overlooked in most treatments of GR. The Poincaré symmetry describes the kinematical content of the theory, while the diff invariance is responsible for the dynamics.

Our considerations in this line of research are somewhat at odds with the traditional approaches to the problem of gauge invariance that has emerged in the last century. GR was formulated as a geometrical theory of the space-time. The GR field equations determine the curvature of the space-time, while the gauge transformations transform the coordinate vectors in the local tangent spaces. This geometric language found its way into the description of the local unitary gauge transformations of the Standard Model. In the Standard Model instead of the tangent bundle we have a unitary fiber bundle, and the local gauge transformations deal with the transformations of the local unitary connections.

There is also an attempt to unify the unitary gauge transformations with GR within a geometrical viewpoint. From this viewpoint, the unitary gauge transformations are associated with the tangent bundle of the compact spatial coordinates. These are the so-called Kaluza-Klein theories. The agreement of these theories with GR and Standard Model equations requires an introduction of compactified spatial dimensions and an additional scalar field (dilaton) [30]. However, there is no simple elegant mechanism that explains the exact nature of the compactification of the extra dimensions.

Another problem with Kaluza-Klein theories is the interpretation of the extra dimensions. In these theories, the radius of compactified dimensions is related to the charges of the loacl gauge fields. In GR geometry is dynamically dependenat, and thus the radius of the extra dimensions should be location-dependant. This means that the charges should be location dependant as well, which is clearly not supported by the experimantal evidence. In order to fix this incongruity, additional fields are needed to stabilize these radii. However, it is not clear if this approach leads to stable solutions.

Historically, the only other functional-theoretic approach to gravitational interaction is that of Feynman [51, 52]. In his lectures and articles, Feynman startes out by introducing the most general interacting spin-2 field that couples to stress-energy tensor. By requiring the consistency of this theory he arrives at the field equations of General Relativity. Feynman did nor explicitly use any gauge principle, but the resulting equations satisfied the property of general covariance. This approach was further pursued and made more rigorous by Robert Wald [53].

1.4 Alternative Theories of Gravity

Over the years there have been many alternative theories of gravity that either compete with or complement GR. This is not entirely surprising since gravity is experimentally least well understood of all physical forces, and there is a considerable room for possible alternative theories. We'll list here a few of better known such theories.

1.4.1 Brans-Dicke theory

This is a scalar-tensor theory of gravity, which means that in addition to the usual tensor (spin-2) field of GR, gravity is propagated with an additional scalar field $\phi(x)$ [54, 55]. The original motivation for this theory was to introduce a position dependent gravitational constant. Just like the GR this theory relies on curved spacetime and geometrical interpretation of gravitational interactions. It has withstood most experimental scrutiny so far, albeit with some restriction on its parameters.

1.4.2 Emergent Gravity

This is an idea that the gravitational degrees of freedom emerge as a macroscopic mean field approximation of the underlying dynamical theory in the same way that the fluid dynamics equations are a macroscopic form of the underlying exact equations of motion. [56, 57]. The idea was originally proposed by Sakharov and it's also known as the induced gravity.

1.4.3 Tensor-Vector-Scalar gravity

Tensor-Vector-Scalar gravity (TeVeS) is a Modified Newtonian dynamics theory of gravity. [58, 59] This theory was motivated by the desire to come up with a modified theory of gravity that can explain the galaxy rotation problem without invoking the dark matter.

1.5 Gravitation and Quantization

Besides the desire for the unification of the gauge interactions, another motivation for our line of research is the goal of quantization of gravitational interaction. GR is the only fundamental physical theory that has not been quantized. So far all known quantization schemes have failed to yield a consistent quantum theory. The only approaches that have been successful are based on an assumption that GR is a low energy limit of some more fundamental underlying theory (String Theory, Loop Quantum Gravity, etc.). All these theories assume some complicated non-local structure at very small length-scales (on the order of the Planck Length).

Another possibility is that gravity is fundamentally a non-quantiziable theory. According to this view, GR is an emergent theory stemming from some yet unknown underlying theory. It would not make sense to quantize GR any more than to quantize the Navier-Stokes equations.

In this paper we adopt the view that we don't need to go beyond the general framework of quantum field theory in order to quantize gravity. We explicitely show how quantization can be achieved with the new model of gravitational interaction that we introduce.

1.6 Organization of this Thesis

We have organized this thesis in the following way:

In Chapter 2 we explore gravity as a non-local gauge field theory and we construct a non-local unitary gauge field theory. We show how we can interpret gravitational interaction as a subset of this most general unitary-theory interaction, and we present some evidence of how such a theory would agree with GR. We also present a case that many other transformations that we regularly encounter in physics can be best understood as a special case of non-local transformations.

In Chapter 3 we present an alternative approach to the introduction of non-local interaction field. We start with a most general translationally-invariant effective ϕ^4 field theory and we show that in a most general case we wold need to introdice a four-point function for the gauge-field propagator if we were to turn such a theory into a gauge field theory. We also make a historical analogy with the development of the renormalizable theory of weak interaction.

Because scalar field theories are much easier to quantize than the corresponding theories with spin degrees of freedom, in Chapter 4 we explore the quantization of the non-local ϕ^4 theories. We introduce a class of models of such interactions that are renormalizable and reduce to gravity-like interaction in the low-energy limit.

In Chapter 5 we deal with the quantization of non-local unitary gauge theory. We show that the perturbative quantization yields non-divergent diagrams.

In Chapter 6 we deal with the gauging of the spin degrees of freedom. We show that in non-local gauge field theories gauging of the spin degrees of freedom makes sense, and we indicate how to go about constructing such a theory. Because of its speculative nature, this chapter is more open-ended than the rest of this thesis.

In the concluding chapter we recapitulate the main aspects of the non-local gauge field theory, why it yields a very natural unification of gravitational interaction and gravity, and we review once again the features that make its quantum theory renormalizable. We also sketch the direction for the future work.

2 Non-Local Gauge Invariance in Physical Theories I: Classical Theory

2.1 Overview

In this chapter we take a closer look at the non-local transformations in physics. Since the primary goal of this work is a deeper understanding of gravity, we focus our attention on understanding gravity as a non-local gauge field theory. However, many other transformations commonly used in physics are also non local (CPT invariance, Fourier transforms, Poincaré and conformal groups) and the last section of this chapter briefly discusses those as well.

Section 2.2 shows that in terms of concepts developed in this thesis, General Relativity is a theory of non-local gauge transformations. Section 2.3 develops the general properties of a non-local unitary gauge field theory. Section 2.4 develops an alternative theory of gravity as a non-local unitary gauge field theory. It shows how this new theory agrees with General Relativity, and combines it with electromagnetic interaction for a new unified theory. Section 2.4 constitutes the bulk of this chapter. Section 2.5 is dedicated to the better understanding of the non-local transformations, and section 2.6 shows that many other transformations commonly used in Physics can also be viewed as non-local transformations.

2.2 General Relativity as a Gauge Field Theory

The gauge invariance is the fundamental organizing principle for both the field theories of the Standard Model and for General Relativity (GR). However, the gauge invariance acts quite differently in the Standard Model than it does in GR. In the Standard Model we locally gauge fields, while in GR we locally gauge the coordinates. Clearly, the gauge invariance has two very different physical meanings. We would like to remedy this somehow. The modern view of the fields that emerged from the Quantum Field Theory (QFT) [62, 63, 65] is that the field configurations are the elements of a vector space, and that the coordinates are just the continuous indices of these vectors. As such they do not have any immediate physical meaning. Indeed, GR can be formulated as a gauge theory of diffeomorphisms of the fields. However, a diffeomorphism transformation of a field is not local transformation (it certainly is not global); it mixes components of the field from different space-time points. It is nevertheless a linear transformation of the field, which is straightforward to see:

Let $f(\mathbf{x})$ and $g(\mathbf{x})$ be two field configurations. Then by the assumption so is $h(\mathbf{x}) = af(\mathbf{x}) + bg(\mathbf{x})$. Let a diffeomorphism act on a field in the following way:

$$f(\mathbf{x}) \rightarrow D_{\mathbf{y}}f(\mathbf{x}) = f(\mathbf{y}(\mathbf{x})), \tag{2.1}$$

where $\mathbf{y} : \mathbb{R} \rightarrow \mathbb{R}$ is a function such that $\det \partial \mathbf{y}/\partial \mathbf{x} \neq 0$, i.e. it is a diffeomorphism.

We see that

$$D_{\mathbf{y}}h(\mathbf{x}) = h(\mathbf{y}(\mathbf{x})) = af(\mathbf{y}(\mathbf{x})) + bg(\mathbf{y}(\mathbf{x})) = aD_{\mathbf{y}}f(\mathbf{x}) + bD_{\mathbf{y}}g(\mathbf{x}), \tag{2.2}$$

and therefore we see that $D_{\mathbf{y}}$ is a linear operator.

What we mean by global, local, and non-local is the following: if fields were vectors in a finite-dimensional vector space, then a global transformation would correspond to a multiplication by a scalar, a local transformation would be a multiplication by a diagonal matrix, and a non-local transformation would be a multiplication by an arbitrary off-diagonal matrix. Using this analogy we can re-interpret what we mean by global, local and non-local transformation in the following way:

Global transformations. In the world of finite-dimensional matrices, a global transformation will correspond to a diagonal matrix with all of its diagonal elements equal:

$$\begin{pmatrix} v_1 \\ v_2 \\ v_3 \end{pmatrix} \rightarrow \begin{pmatrix} a & 0 & 0 \\ 0 & a & 0 \\ 0 & 0 & a \end{pmatrix} \begin{pmatrix} v_1 \\ v_2 \\ v_3 \end{pmatrix} = \begin{pmatrix} av_1 \\ av_2 \\ av_3 \end{pmatrix}$$

Local transformations. Local transformations will correspond to an arbitrary diagonal matrix:

$$\begin{pmatrix} v_1 \\ v_2 \\ v_3 \end{pmatrix} \rightarrow \begin{pmatrix} a_1 & 0 & 0 \\ 0 & a_2 & 0 \\ 0 & 0 & a_3 \end{pmatrix} \begin{pmatrix} v_1 \\ v_2 \\ v_3 \end{pmatrix} = \begin{pmatrix} a_1 v_1 \\ a_2 v_2 \\ a_3 v_3 \end{pmatrix}$$

Non-local transformations. Non-local transformations correspond to an arbitrary matrix:

$$\begin{pmatrix} v_1 \\ v_2 \\ v_3 \end{pmatrix} \rightarrow \begin{pmatrix} a_{11} & a_{12} & a_{13} \\ a_{21} & a_{22} & a_{23} \\ a_{31} & a_{32} & a_{33} \end{pmatrix} \begin{pmatrix} v_1 \\ v_2 \\ v_3 \end{pmatrix} = \begin{pmatrix} a_{11}v_1 + a_{12}v_2 + a_{13}v_3 \\ a_{21}v_1 + a_{22}v_2 + a_{23}v_3 \\ a_{31}v_1 + a_{32}a_2 + a_{33}v_3 \end{pmatrix}$$

It is clear from the last line that the non-local transformations will mix field components from different spacetime points.

The foregoing discussion suggests two things. First, we see that it is not necessary for a linear transformation of a field to be local in order for the corresponding gauge field theory to be considered a viable dynamical physical theory. We are thus led to consider the non-local linear unitary transformations of the Dirac field in the Standard Model. Furthermore, we would like to reformulate the gravitational theory as a theory of gauge-invariant fields in a manner that is completely analogous to the non-local unitary gauge theory.

2.3 Non-Local Unitary Gauge Theory

2.3.1 General Properties

Although the author has independently arrived at the results presented in this section, the Non-Local Unitary Gauge Theory has been considered in the past literature [21]. We will follow our own derivation.

The action for the Dirac field is given by

$$S = \int \left(\bar{\psi}(x) i\,\slashed{\partial} \psi(x) + m \bar{\psi}(x) \psi(x) \right) d^d x. \qquad (2.3)$$

The mass term in the action for the Dirac field, $\int m \bar{\psi}(x) \psi(x) d^d x$, is obviously invariant if we multiply the field ψ by an arbitrary phase, $\psi \to e^{i\theta} \psi$. If, in addition, we require that the phase factor be a function of position, $\theta(x)$, then we are forced to introduce the covariant derivative $\mathcal{D}^\mu = \partial^\mu - i A^\mu(x)$ that insures that the kinematic term is also gauge invariant. This is accomplished by introducing a gauge invariant action for the field strength $A^\mu(x)$. It is easy to check that the field strength defined by

$$F^{\mu\nu} = [\mathcal{D}^\mu, \mathcal{D}^\nu] = \partial^\mu A^\nu(x) - \partial^\nu A^\mu(x) \qquad (2.4)$$

will be invariant under the change of gauge

$$A^\mu(x) \to A^\mu(x) + \partial^\mu \lambda(x), \qquad (2.5)$$

where $\lambda(x)$ is an arbitrary function.

If we now write the mass term as $\int m \bar{\psi}(y) \delta(y - x) \psi(x) d^d x d^d y$, then this suggests that we can treat the entire field ψ as a vector in a functional vector space for which the mass term provides an inner product. In general, the inner product may not be positive-definite, as is the case for instance with Dirac spinor. This inner product is invariant under an arbitrary unitary transformation of $\psi(x)$, that is, under the transformation

$$\psi(x) \to \int U(x, y) \psi(y) d^d y, \qquad (2.6)$$

such that

$$\int U^*(x, z) U(z, y) d^d z = \delta(x - y), \qquad (2.7)$$

or in the operator language

$$\psi \to \hat{U} \psi, \quad \text{with} \quad \hat{U}^\dagger \hat{U} = \hat{I}. \qquad (2.8)$$

The most general gauge invariant derivative that would be consistent with this sort of gauge transformation would be a general hermitian operator $\hat{\mathcal{D}}^\mu$, but in order to make the connection with the local case more transparent, it is advantageous to make the distinction between the derivative term and the rest of the operator. Thus we would write

$$\hat{\mathcal{D}}^\mu = \partial^\mu - i\hat{A}^\mu. \tag{2.9}$$

The gauge transformation for the invariant derivative is

$$\hat{\mathcal{D}}^\mu \to \hat{U}\hat{\mathcal{D}}^\mu\hat{U}^\dagger. \tag{2.10}$$

The corresponding field strength

$$\hat{F}^{\mu\nu} = [\hat{\mathcal{D}}^\mu, \hat{\mathcal{D}}^\nu], \tag{2.11}$$

is not invariant under the general unitary transformation. Instead, it transforms as

$$\hat{F}^{\mu\nu} \to \hat{U}\hat{F}^{\mu\nu}\hat{U}^\dagger. \tag{2.12}$$

On the other hand, the Lorentz invariant action

$$S = \frac{1}{4}tr(\hat{F}^{\mu\nu}\hat{F}_{\mu\nu}) \tag{2.13}$$

remains unchanged.

Since we are unable to employ the Leibniz rule when computing the action of the derivative on non-local operator $\hat{U}$, it is not clear at first sight how to write the transformation of the operator $\hat{A}^\mu$ by itself. We know that it should follow directly from the transformation properties of $\hat{\mathcal{D}}^\mu$ that was presented by the equation (2.10), but we will first concentrate on an infinitesimal unitary gauge transformation before presenting the transformations we are looking for. The transformation that we are looking for will be:

$$\hat{U} = \hat{I} + i\hat{\theta}, \quad \hat{U}^\dagger = \hat{I} - i\hat{\theta}, \tag{2.14}$$

where $\hat{\theta}$ is a hermitian operator, $\hat{\theta} = \hat{\theta}^\dagger$. Under this kind of infinitesimal transformation the partial derivative transforms as (neglecting the second order terms):

$$\partial^\mu \to \hat{U}\partial^\mu\hat{U}^\dagger = (\hat{I} + i\hat{\theta})\partial^\mu(\hat{I} - i\hat{\theta}) = \partial^\mu + i[\hat{\theta}, \partial^\mu]. \tag{2.15}$$

Similarly, the operator $\hat{A}^\mu$ transforms as

$$\hat{A}^\mu \to \hat{U}\hat{A}^\mu\hat{U}^\dagger = (\hat{I} + i\hat{\theta})\hat{A}^\mu(\hat{I} - i\hat{\theta}) = \hat{A}^\mu + i[\hat{\theta}, \hat{A}^\mu], \qquad (2.16)$$

that is, the covariant partial derivative as a whole transforms as

$$\hat{\mathcal{D}}^\mu = \hat{\mathcal{D}}^\mu + i[\hat{\theta}, \hat{\mathcal{D}}^\mu] = \partial^\mu + i\hat{A}'^\mu, \qquad (2.17)$$

where

$$\hat{A}'^\mu = \hat{A}^\mu + [\hat{\theta}, \partial^\mu,] + [\hat{\theta}, \hat{A}^\mu]. \qquad (2.18)$$

From this we infer that the gauge transformation that is exact to all orders will be

$$\hat{\mathcal{D}}^\mu \to \exp([i\hat{\theta}, \ ])\hat{\mathcal{D}}^\mu, \qquad (2.19)$$

which is essentially the Baker-Campbell-Hausdorff formula. This transformation law also means that the covariant derivative $\hat{\mathcal{D}}^\mu$ transforms in the adjoint representation of the non-local unitary group. The field potential then transforms as

$$\hat{A}^\mu \to \exp([i\hat{\theta}, \ ])\hat{\mathcal{D}}^\mu - \partial^\mu. \qquad (2.20)$$

Most of the work so far has been fairly formal, and the same general framework had been developed independently by Z. Dongpei [21]. In the next few section we will attempt to make non-local gauge field theory more amenable to calculations and more relevant to the other physical theories.

2.3.2 A More Fundamental Derivation of the Gauge Invariant Derivative

Although the definition of the covariant derivative given in the previous section is correct, we would like to have a more controlled way of deriving it. This would particularly be useful when the gauge group does not turn out to be unitary, as we shall see in the next section.

In local gauge theories a useful quantity in definition of the gauge invariant derivative is the comparator $C(x, y)$, a bi-local function that contains the information of how two phases are related at different points. This is the basis of definition of the covariant derivative in terms of the standard parallel transport. It helps to compare two values of a field along a given geodesic. The definition of the gauge derivative then becomes

$$n^\mu \mathcal{D}_\mu \psi = \lim_{\epsilon \to 0} \frac{1}{\epsilon} [\psi(x + \epsilon n) - C(x + \epsilon n, x)\psi(x)], \qquad (2.21)$$

where the comparator has the following transformation law

$$C(x, y) \to e^{\theta(y)} C(y, x) e^{-\theta(x)}. \qquad (2.22)$$

.

For a non-local gauge transformation we cannot introduce such a comparator because we need to compare the field at one point to the field at all other points. Thus, it is not straightforward to compare $\psi(x + \epsilon n)$ with $\psi(x)$. In a sense, we need to compare a value of the field ψ at one spacetime point with a value at another along all the possible paths between those two points. This, however, is only a loose analog because the geometrical analogy breaks down at this point.

Instead, we define an infinitesimal transformation of the field $\psi(x)$ as a general unitary transformation $\hat{\delta}_\epsilon$, where $\hat{\delta}_\epsilon \to \hat{I}$ as $\epsilon \to 0$. From this point of view, $\hat{\delta}_\epsilon$ is a generator of unitary transformation. The definition of the covariant derivative is then given by

$$n^\mu \mathcal{D}_\mu \psi = \lim_{\epsilon \to 0} \frac{1}{\epsilon} [\hat{\delta}_{\epsilon \hat{n}} \psi(x) - \psi(x)]. \qquad (2.23)$$

where n^μ is an arbitrary unit four-vector.

We see that we recover the previous definition of the covariant derivative if we take $\delta_{\epsilon \mathbf{n}}$ to be

$$\delta_{\epsilon \mathbf{n}} = e^{-\epsilon n^\mu \partial_\mu - i \epsilon n^\mu \hat{A}_\mu}. \qquad (2.24)$$

2.4   The Gauging of Diffeomorphisms

At this point we would like to take another look at General Relativity. We assume that the physical content of the theory is the gauge group of diffeomorphisms of the field, and not the diffeomorphisms of manifolds. This raises two questions. First, the mass term in the Dirac action is not invariant if we just perform a linear diffeomorphism of the field. We would need to introduce a dynamical measure that absorbs this effect of the diffeomorphisms. This dynamical measure would mean that the Jacobian of the corresponding integrals would be field-valued as well, and invariant under the gauge transfromations. We leave this issue aside for now.

The second question deals with the covariant differentiation. We would like to construct a covariant derivative $\mathcal{D}^\mu$ such that when the field $\psi(x)$ goes to $\hat{G}\psi(x)$, where $\hat{G}$ is a general diffeomorphism, then the derivative $\mathcal{D}^\mu$ goes to $\hat{G}\mathcal{D}^\mu\hat{G}^{-1}$.

We will start defining the diffeomorphism-invariant derivative by considering the action of infinitesimal diffeomorphisms on fields. We have

$$\psi(x^\mu) \to \psi(x^\mu + \epsilon h^\mu(x^\nu)) \approx \psi(x^\mu) + \epsilon h^\sigma(x^\mu)\partial_\sigma\psi(x^\mu), \tag{2.25}$$

so that

$$\hat{G} \approx \hat{I} + h^\sigma(x^\mu)\partial_\sigma, \quad \hat{G}^{-1} \approx \hat{I} - h^\sigma(x^\mu)\partial_\sigma. \tag{2.26}$$

Under this transformation the partial derivative transforms as

$$\partial^\mu \to \hat{G}\partial^\mu\hat{G}^{-1} \approx (\hat{I} + h^\sigma\partial_\sigma)\partial^\mu(\hat{I} - h^\sigma\partial_\sigma) \approx \partial^\mu + [h^\sigma\partial_\sigma, \partial^\mu] = \partial^\mu - \partial^\mu h^\sigma\partial_\sigma, \tag{2.27}$$

where we have again neglected the second order terms in h^σ.

We see that the partial derivatives do not commute with diffeomorphisms. In order to correct that, we need to introduce an additional gauge field with which we can construct a covariant derivative $\mathcal{D}^\mu$. A gauge field that will accomplish the desired invariance needs to be a derivative-valued field. From these considerations we obtain the explicit expression $\mathcal{D}^\mu$

$$\mathcal{D}^\mu = \partial^\mu + f^{\mu\nu}\partial_\nu. \tag{2.28}$$

$f^{\mu\nu}\partial_\nu$ will transform as

$$f^{\mu\nu}\partial'_\nu = f^{\mu\nu}\partial_\nu + [h^\sigma\partial_\sigma, \partial^\mu] + [h^\sigma\partial_\sigma, f^{\mu\nu}\partial_\nu], \tag{2.29}$$

so that

$$f^{\mu\nu'} = f^{\mu\nu} - \partial^\mu h^\nu - f^{\mu\sigma}\partial_\sigma h^\nu + h^\sigma\partial_\sigma f^{\mu\nu}. \tag{2.30}$$

As it stands, $i\mathcal{D}^\mu$ is not hermitian. In order to make it hermitian we require that

$$\partial_\nu f^{\mu\nu} = 0. \tag{2.31}$$

An interesting feature of this covariant derivative is that it does not depend on the spin of the field on which it acts. It has the same form for scalars, spinors, vector fields, etc. This, for instance, is not the case in GR. This is understandable since the group of diffeomorphisms acts on the coordinates of the field and not on its indices.

We will also write down the equation of motion for the scalar field. This is of interest for the sake of comparison with the same equation in GR. The gauge invariant equation of motion is given as

$$
\begin{aligned}
0 &= \mathcal{D}^\mu \mathcal{D}_\mu \phi(x) \\
&= (\partial^\mu + f^\mu_\nu \partial^\nu)(\partial_\mu + f^\nu_\mu \partial_\nu)\phi(x) \\
&= (\eta^{\mu\nu} + 2f^{\mu\nu} + f^{\gamma\mu} f^\nu_\gamma)\partial_\mu \partial_\nu \phi(x) + (\partial_\mu f^{\mu\nu} + f_{\mu\beta}\partial^\beta f^{\mu\nu})\partial_\nu \phi(x).
\end{aligned} \tag{2.32}
$$

In the GR the equation of motion of the scalar field is obtained in the following way ([24]): the first partial derivative of the scalar field is left as it is in the non-GR case. That is, we still write $\partial_\mu \phi(x)$. This is because in GR covariant derivatives are the same for scalar fields as the ordinary partial derivatives. The second partial derivative, however, will act on a Lorentzian vector field $\partial_\mu \phi(x)$ and we have

$$\partial^\mu \partial_\mu \phi(x) + \Gamma^\nu_{\alpha\alpha} \partial_\nu \phi(x) = 0, \tag{2.33}$$

where all the contractions have been made with the use of GR metric $g^{\mu\nu}(x)$, and $\Gamma^\mu_{\nu\rho}(x)$ are known in GR as Christoffel symbols or connection. In GR $\Gamma^\mu_{\nu\rho}(x)$ is not a tensor.

We see that the scalar field equation assumes the same form in non-local gauge field theory as the scalar field equation in GR, and therefore it is possible to interpret the predictions of one in terms of the other one. However, this would require a comprehensive tests of GR in the strong field/wave mechanics limit. These kinds of tests are not viable with the present day experiments. The best tests of GR right now are those that involve the geodesic motion of relativistic particles. We would need to show that the characteristics of the scalar field equation (2.32) obey the geodesic equation, at least to the first order in the wavelength to radius of curvature ratio.

2.4.1   Particle Motion Equation

In order to make a connection between our theory and GR we will use the formalism of the wave mechanics, and derive the classical equations of motion for a relativistic particle. We need to make the following substitution:

$$-i\partial_\mu \to p_\mu. \tag{2.34}$$

Then we need to show that the classical particle dynamics equation reduces to the geodesic equation.

In order to motivate our argument we shall first describe the motion of the relativistic particle in an electromagnetic field. The classical action is given by

$$S = -\int \left[m\sqrt{\eta^{\mu\nu}\frac{dx^\mu}{ds}\frac{dx^\nu}{ds}} + \frac{dx_\mu}{ds}A^\mu(x) \right] ds. \tag{2.35}$$

Hamilton's principle yields the Euler-Lagrange equations,

$$\frac{d}{ds}\left(\frac{\partial L}{\partial \left(\frac{dx_\mu}{ds} \right)} \right) - \partial^\mu L = 0, \tag{2.36}$$

where the Lagrangian is

$$L = -\left[m\sqrt{\eta^{\mu\nu}\frac{dx^\mu}{ds}\frac{dx^\nu}{ds}} + \frac{dx_\mu}{ds}A^\mu(x) \right]. \tag{2.37}$$

The Euler-Lagrange equations give us

$$m\frac{d^2x^\mu}{ds^2} = (\partial^\mu A^\nu - \partial^\nu A^\mu)\frac{dx_\nu}{ds} = F^{\mu\nu}\frac{dx_\nu}{ds}. \tag{2.38}$$

The transition to the Hamiltonian view is straightforward, although it suffers from some interpretive ambiguities. The conjugate momentum 4-vector is defined by

$$p^\mu = -\frac{\partial L}{\partial \left(\frac{dx_\mu}{ds} \right)} = mu^\mu + A^\mu. \tag{2.39}$$

A Hamiltonian can be defined by

$$H = \frac{1}{2}(p_\mu u^\mu - L),$$ (2.40)

which leads to

$$H = \frac{1}{2m}(p_\mu - A_\mu)(p^\mu - A^\mu) - \frac{1}{2}m.$$ (2.41)

Hamilton's equations are

$$\frac{dx^\mu}{ds} = \frac{\partial H}{\partial p_\mu} = \frac{1}{m}(p^\mu - A^\mu),$$ (2.42)

and

$$\frac{dp^\mu}{ds} = -\frac{\partial H}{\partial x_\mu} = \frac{1}{m}(p_\nu - A_\nu)\partial^\mu A^\nu.$$ (2.43)

These equations can be immediately shown to be equivalent to the Euler-Lagrange equations.

Although the above Hamiltonian has several problems (it is a Lorentz scalar, it identically vanishes), it has proven more convenient for the description of the motion of a relativistic particle under the influence of force that arises due to the diffeomorphism invariance. We obtain the corresponding equation of motion under the influence of gravitational force by making the substitution $A^\mu \to f^{\mu\nu}\partial_\nu$. The equivalent of the equation (2.41) would be

$$\begin{aligned} H &= \frac{1}{2m}(p_\mu - f_{\mu\nu}p^\nu)(p^\mu - f^{\mu\nu}p_\nu) - \frac{m}{2} \\ &= \frac{1}{2m}g^{\mu\nu}p_\mu p_\nu - \frac{m}{2}, \end{aligned}$$ (2.44)

where $g^{\mu\nu} = \eta^{\mu\nu} - [f^{\mu\nu} + f^{\nu\mu} - 1/2(f^{\mu\alpha}f^\nu_\alpha - f^{\nu\alpha}f^\mu_\alpha)]$. The Hamiltonian equations are

$$\frac{dx^\mu}{ds} = \frac{\partial H}{\partial p_\mu} = \frac{1}{m}g^{\mu\nu}p_\nu$$ (2.45)

and

$$\frac{dp^\mu}{ds} = -\frac{\partial H}{\partial x_\mu} = -\frac{1}{2m}\partial^\mu g^{\alpha\beta}p_\alpha p_\beta.$$ (2.46)

These equations yield

$$\begin{aligned}
\frac{d^2 x^\mu}{ds^2} &= \frac{1}{m} \frac{dg^{\mu\nu}}{ds} p_\nu + \frac{1}{m} g^{\mu\nu} \frac{dp_\nu}{ds} \\
&= \frac{1}{m} \frac{dx^\alpha}{ds} \partial_\alpha g^{\mu\nu} p_\nu + \frac{1}{m} g^{\mu\nu} \left(-\frac{1}{2m} \partial_\nu g^{\alpha\beta} p_\alpha p_\beta \right) \\
&= \frac{dx^\alpha}{ds} \partial_\alpha g^{\mu\nu} \tilde{g}_{\nu\beta} \frac{dx^\beta}{ds} - \frac{1}{2} g^{\mu\nu} \left(\partial_\nu g^{\alpha\beta} \tilde{g}_{\alpha\rho} \tilde{g}_{\beta\sigma} \frac{dx^\rho}{ds} \frac{dx^\sigma}{ds} \right) \\
&= \Gamma^\mu_{\ \rho\sigma} \frac{dx^\rho}{ds} \frac{dx^\sigma}{ds},
\end{aligned} \tag{2.47}$$

where

$$\begin{aligned}
\Gamma^\mu_{\ \rho\sigma} &= \frac{1}{2} \left(\partial_\rho g^{\mu\beta} \tilde{g}_{\beta\sigma} + \partial_\sigma g^{\mu\beta} \tilde{g}_{\beta\rho} - g^{\mu\nu} \partial_\nu g^{\alpha\beta} \tilde{g}_{\alpha\rho} \tilde{g}_{\beta\sigma} \right) \\
&= \frac{1}{2} g^{\mu\nu} \left(\partial_\rho g^{\alpha\beta} \tilde{g}_{\beta\sigma} \tilde{g}_{\nu\alpha} + \partial_\sigma g^{\alpha\beta} \tilde{g}_{\beta\rho} \tilde{g}_{\nu\alpha} - \partial_\nu g^{\alpha\beta} \tilde{g}_{\alpha\rho} \tilde{g}_{\beta\sigma} \right)
\end{aligned} \tag{2.48}$$

where $\tilde{g}^{\mu\nu}$ is the matrix inverse of $g_{\mu\nu}$, that is

$$\tilde{g}^{\mu\nu} g_{\nu\alpha} = \delta^\mu_\alpha, \tag{2.49}$$

and where the corresponding index contaction is obtained with respect to the flat metric $\eta^{\mu\nu}$.

We see that we have obtained the geodesic equation for the motion of a relativistic particle in a gravitational field.

The Lagrangian that corresponds to the equation (39) is

$$L = g^{\mu\nu} p_\mu p_\nu, \tag{2.50}$$

so that the total Lagrangian for Electromagnetic and Gravitational fields is

$$L = p_\mu A^\mu + p_\mu g^{\mu\nu} p_\nu. \tag{2.51}$$

The usual approach to deriving the equation of motion for a particle in gravitational field in GR is to make an a-priori assumption that particle moves along a geodesic. In that case the equation of motion is just the equation for the geodesic in a given geometry. In this section we have demonstrated that in non-local gauge theory it is possible to end up with a geodesic equation without any a-priori assumptions about the nature of the particle's motion. Thus, the geometrical interpretation of the gravitational interaction is just one of the possible interpretations, and not necessarily the only one.

2.4.2 Field Equations

Let us look at the following expression:

$$[f^\nu \partial_\nu, g^\beta \partial_\beta] = (f^\nu \partial_\nu g^\beta - g^\nu \partial_\nu f^\beta)\partial_\beta. \tag{2.52}$$

We see that the Lie algebra generated by operators of the form $f^\nu \partial_\nu$ closes. These are anti-hermitian operators, provided that $\partial_\mu f^\nu = 0$, and therefore the group whose elements are of the form $\exp(f^\nu \partial_\nu)$ is a subgroup of the general non-local unitary group. This is the group of "local translations," and we see that it has the same Lie algebra as the group of general diffeomorphisms. Therefore these two groups "locally" look the same, and it is quite possible that the true gauge group of the gravitational interaction is in fact this subgroup of the general Non Local unitary group. This point of view has several conceptual advantages. First, we would not need to introduce any new dynamical measure for the Action. Furthermore, we would then only need one unifying gauge principle for all the interactions.

The general field equations in the Standard Model take the form

$$[\hat{\mathcal{D}}^{\mu\mathrm{tr}}, [\hat{\mathcal{D}}_\mu, \hat{\mathcal{D}}_\nu]]_{ab} = -g\bar{\psi}(x)_a \gamma_\mu \psi(x)_b, \tag{2.53}$$

that is, the source of the field is just the local current, and the left hand side is a matrix.

We can take the left hand side of the above equation to be valid even in the non-local theory. However, in such a situation, the source term becomes a bit more problematic. For now we shall only pay attention to the field equations in vacuum. We have

$$[\hat{\mathcal{D}}^{\mu\mathrm{tr}}, [\hat{\mathcal{D}}_\mu, \hat{\mathcal{D}}_\nu]] = 0, \tag{2.54}$$

If we presently consider only the gravitational interaction, then we have

$$\hat{\mathcal{D}}^{\mu} = \partial^{\mu} + f^{\mu\alpha}\partial_{\alpha},$$

$$\begin{aligned}
\hat{F}^{\mu\nu} = [\hat{\mathcal{D}}_{\mu}, \hat{\mathcal{D}}_{\nu}] &= [\partial^{\mu} + f^{\mu\alpha}\partial_{\alpha}, \partial^{\nu} + f^{\nu\beta}\partial_{\beta}] \\
&= \partial^{\mu}f^{\nu\beta}\partial_{\beta} - \partial^{\nu}f^{\mu\alpha}\partial_{\alpha} + f^{\mu\alpha}\partial_{\alpha}f^{\nu\beta}\partial_{\beta} - f^{\nu\beta}\partial_{\beta}f^{\mu\alpha}\partial_{\alpha} \quad (2.55) \\
&= (\partial^{\mu}f^{\nu\alpha} - \partial^{\nu}f^{\mu\alpha} + f^{\mu\beta}\partial_{\beta}f^{\nu\alpha} - f^{\nu\beta}\partial_{\beta}f^{\mu\alpha})\partial_{\alpha} \\
&= (F^{\mu\nu\alpha} + G^{\mu\nu\alpha})\partial_{\alpha},
\end{aligned}$$

where now

$$\begin{aligned}
F^{\mu\nu\alpha} &= \partial^{\mu}f^{\nu\alpha} - \partial^{\nu}f^{\mu\alpha}, \\
G^{\mu\nu\alpha} &= f^{\mu\beta}\partial_{\beta}f^{\nu\alpha} - f^{\nu\beta}\partial_{\beta}f^{\mu\alpha}.
\end{aligned} \qquad (2.56)$$

We see that $F^{\mu\nu\alpha}$ is analogous to the electromagnetic field potential $F^{\mu\nu}$, while $G^{\mu\nu\alpha}$ is a nonlinear term that arises due to noncommutativity of the field $f^{\mu\beta}\partial_{\beta}$.

Finally we have

$$\begin{aligned}
[\hat{\mathcal{D}}^{\mu\mathrm{tr}}, [\hat{\mathcal{D}}_{\mu}, \hat{\mathcal{D}}_{\nu}]] &= [\hat{\mathcal{D}}^{\mu\mathrm{tr}}, \hat{F}^{\mu\nu}] \\
&= [\partial_{\mu} + f_{\mu\gamma}\partial^{\gamma}, F^{\mu\nu\alpha}\partial_{\alpha} + G^{\mu\nu\alpha}\partial_{\alpha}] \\
&= [\partial_{\mu}, F^{\mu\nu\alpha}\partial_{\alpha}] + [\partial_{\mu}, G^{\mu\nu\alpha}\partial_{\alpha}] + [f_{\mu\gamma}\partial^{\gamma}, F^{\mu\nu\alpha}\partial_{\alpha}] + [f_{\mu\gamma}\partial^{\gamma}, G^{\mu\nu\alpha}\partial_{\alpha}] \\
&= (\partial_{\mu}F^{\mu\nu\alpha} + \partial_{\mu}G^{\mu\nu\alpha})\partial_{\alpha} + f_{\mu\gamma}\partial^{\gamma}F^{\mu\nu\alpha}\partial_{\alpha} - F^{\mu\nu\alpha}\partial_{\alpha}f_{\mu\gamma}\partial^{\gamma} \\
&\quad + f_{\mu\gamma}\partial^{\gamma}G^{\mu\nu\alpha}\partial_{\alpha} - G^{\mu\nu\alpha}\partial_{\alpha}f_{\mu\gamma}\partial^{\gamma} \\
&= (\partial_{\mu}F^{\mu\nu\alpha} + \partial_{\mu}G^{\mu\nu\alpha} + f_{\mu\gamma}\partial^{\gamma}F^{\mu\nu\alpha} - F^{\mu\nu\gamma}\partial_{\gamma}f_{\mu}^{\alpha} \\
&\quad + f_{\mu\gamma}\partial^{\gamma}G^{\mu\nu\alpha} - G^{\mu\nu\gamma}\partial_{\gamma}f_{\mu}^{\alpha})\partial_{\alpha}.
\end{aligned}$$

$$(2.57)$$

We see from the above that the expression in the parenthesis is second order in derivatives. Therefore, the field equations for $f^{\mu\nu}$ are viable classical field equations. The linear term

$$\partial_{\mu}F^{\mu\nu\alpha} = \partial_{\mu}\partial^{\mu}f^{\nu\alpha} - \partial_{\mu}\partial^{\nu}f^{\mu\alpha} \qquad (2.58)$$

is invariant under $f^{\mu\nu} \to f^{\mu\nu} + \partial^{\mu}\lambda^{\nu}$ which is analogous to the gauge transformation of the electromagnetic field. In the Lorentz gauge $\partial^{\mu}f_{\mu\nu} = 0$ the linearized field equations for the gravitational field become

$$\Box f^{\mu\nu} = 0. \tag{2.59}$$

If the field potential $f^{\mu\nu}$ is time-dependant, then the above linearized form of the field equations give us gravitational waves. Gravitational waves are another prediction of General Relativity, and thus we see that this alternative theory of gravity agrees with GR on this point as well.

In the Weak-Field Approximation in GR we expand the metric tensor around the Minkowski metric $\eta_{\mu\nu}$:

$$g_{\mu\nu} = \eta_{\mu\nu} + h_{\mu\nu} \tag{2.60}$$

The vacuum Einstein Field Equations in the so-called harmonic coordinate system then become

$$\Box h_{\mu\nu} = 0. \tag{2.61}$$

In terms of $f_{\mu\nu}$ the above equation reads

$$\Box(f_{\mu\nu} + f_{\nu\mu}) = 0. \tag{2.62}$$

Since μ and ν are dummy indices, this equation is implied by the gravitational field equation. We see that in the linearized regime the field equations of GR and our gravitational equations are consistent with each other. This is important because so far GR has experimentally only been tested in the linearized regime [34]. Therefore any gravitational theory that reproduces the same results as GR in the linearized regime is physically an equally viable theory of gravitational interaction. It should be noted that, just as in GR, the field equations in this theory are also cubic in field strengths.

2.4.3 Interacting Electromagnetic and Gravitational fields

One of the earliest confirmations of the General relativity came from the observation of deflection of the rays of starlight by the Sun (See [35] for instance). Any alternative theory of gravity would need to agree with these experiments. We have seen before that by redefining gravity as a non-local gauge field theory we recover the geodesic equation exactly. From there we can conclude that any particle that travels under an influence of gravitational field will have to obey this equation as well. However, we would still like to see that the interaction of the gravitational field with light follows naturally from the fundamental equations of the theory. In order to achieve this naturally within the framework of the non-local gauge theory, we need to incorporate both the local gauge transformations and the non-local ones as part of one general transformation.

We will start by considering a non-local unitary group of operators of the form $\exp(i\hat{\theta})$, where $\hat{\theta}$ is a Hermitian operator of the form

$$\hat{\theta} = \theta(x) + i\theta^{\mu}(x)\partial_{\mu}, \tag{2.63}$$

with $\partial_{\mu}\theta^{\mu}(x) = 0$. This constitutes a group that is generated by what we can loosely call local gauge transformations and the "local translations". The first of the two transformations corresponds to the transformations that are related to the electromagnetic field, while the second corresponds to the gauge transformations of the gravitational field. In order to show that the above transformations constitute a group, we will show that they are closed under the Lie bracket:

$$
\begin{aligned}
[\theta_1 + i\theta_1^{\mu}\partial_{\mu}, \theta_2 + i\theta_2^{\nu}\partial_{\nu}] &= i(\theta_1^{\mu}\partial_{\mu}\theta_2 - \theta_2^{\mu}\partial_{\mu}\theta_1) + (\theta_2^{\mu}\partial_{\mu}\theta_1^{\nu} - \theta_1^{\mu}\partial_{\mu}\theta_2^{\nu})\partial_{\nu} \\
&= i(\theta_3 + \theta_3^{\nu}\partial_{\nu}).
\end{aligned}
\tag{2.64}
$$

Thus we see that these transformations indeed form a group. We also make an interesting observation: within this group the subgroup of local transformations is a normal subgroup. We can see this easily by setting $\theta_1^{\mu} = 0$. In that case $\theta_3^{\nu} = 0$ and $\theta_3 = -\theta_2^{\nu}\partial_{\nu}\theta_1$.

Since the standard partial derivatives do not commute with the generators of this group, we need to introduce a covariant derivative that will be able to absorb all the non-commuting terms. This is easily achieved by the introduction of gauge fields. In this case we'll need two gauge fields: $A^{\mu}(x)$ for the local gauge transformations and $A^{\mu\nu}(x)\partial_{\nu}$ for the "local translations". We have replaced $f^{\mu\nu}(x)$ of the preceding chapters with $A^{\mu\nu}(x)$ in order to emphasize the conceptual connection between the local electromagnetic field $A^{\mu}(x)$ and the non-local gravitational field $A^{\mu\nu}(x)\partial_{\nu}$. The covariant derivative thus has the form

$$\mathcal{D}^\mu = \partial^\mu + iA^\mu(x) + A^{\mu\nu}(x)\partial_\nu. \tag{2.65}$$

We obtain the linearized form of the transformation of this new field strength from the equation (2.18). We get

$$A'^\mu \approx A^\mu + \partial^\mu\theta + A^{\mu\alpha}\partial_\alpha\theta + \theta^\nu\partial_\nu A^\mu,$$
$$A'^{\mu\nu} \approx A^{\mu\nu} + \partial^\mu\theta^\nu + \theta^\alpha\partial_\alpha A^{\mu\nu} - A^{\mu\alpha}\partial_\alpha\theta^\nu. \tag{2.66}$$

The field strength associated with this covariant derivative is given by

$$\begin{aligned}
\hat{F}^{\mu\nu} &= [\mathcal{D}^\mu, \mathcal{D}^\nu] \\
&= [\partial^\mu + iA^\mu(x) + A^{\mu\alpha}(x)\partial_\alpha, \partial^\nu + iA^\nu(x) + A^{\nu\beta}(x)\partial_\beta] \\
&= [\partial^\mu, iA^\nu] + [\partial^\mu, A^{\nu\beta}\partial_\beta] + [iA^\mu, \partial^\nu] \\
&\quad + [iA^\mu, A^{\nu\beta}\partial_\beta] + [A^{\mu\alpha}\partial_\alpha, \partial^\nu] + [A^{\mu\alpha}\partial_\alpha, iA^\nu] + [A^{\mu\alpha}\partial_\alpha, A^{\nu\beta}\partial_\beta] \\
&= i(\partial^\mu A^\nu - \partial^\nu A^\mu) + i(A^{\mu\beta}\partial_\beta A^\nu - A^{\nu\beta}\partial_\beta A^\mu) \\
&\quad + (\partial^\mu A^{\nu\alpha} - \partial^\nu A^{\mu\alpha})\partial_\alpha + (A^{\mu\beta}\partial_\beta A^{\nu\alpha} - A^{\nu\beta}\partial_\beta A^{\mu\alpha})\partial_\alpha \\
&= iF^{\mu\nu} + iG^{\mu\nu} + F^{\mu\nu\alpha}\partial_\alpha + G^{\mu\nu\alpha}\partial_\alpha.
\end{aligned} \tag{2.67}$$

$F^{\mu\nu}$ is the usual electromagnetic field strength tensor, and $F^{\mu\nu\alpha}$ is its equivalent for the gravitational field. The nonlinear terms $G^{\mu\nu}$ and $G^{\mu\nu\alpha}$ arise due to the electromagnetic-gravitational interaction and gravitational self-interacting respectively.

The field equation have the form

$$[\mathcal{D}_\nu^{\text{tr}}, \hat{F}^{\mu\nu}] = 0, \tag{2.68}$$

where now

$$\mathcal{D}_\nu^{\text{tr}} = \partial^\mu - iA^\mu(x) + A^{\mu\alpha}(x).\partial_\alpha$$

The field equations now read

$$
\begin{aligned}
[\mathcal{D}^{\mathrm{tr}}_\nu, \hat{F}^{\mu\nu}] &= [\partial^\mu - iA^\mu + A^{\mu\alpha}, \hat{F}^{\mu\nu}] \\
&= [\partial_\nu, \hat{F}^{\mu\nu}] - i[A_\nu, \hat{F}^{\mu\nu}] + [A_{\nu\gamma}\partial^\gamma, \hat{F}^{\mu\nu}] \\
&= i[\partial_\nu, F^{\mu\nu}] + i[\partial_\nu, G^{\mu\nu}] + [\partial_\nu, F^{\mu\nu\alpha}\partial_\alpha] + [\partial_\nu, G^{\mu\nu\alpha}\partial_\alpha] + [A_\nu, F^{\mu\nu}] + [A_\nu, G^{\mu\nu}] \\
&\quad - i[A_\nu, F^{\mu\nu\alpha}\partial_\alpha] - i[A_\nu, G^{\mu\nu\alpha}\partial_\alpha] + i[A_{\nu\gamma}\partial^\gamma, F^{\mu\nu}] \\
&\quad + i[A_{\nu\gamma}\partial^\gamma, G^{\mu\nu}] + [A_{\nu\gamma}\partial^\gamma, F^{\mu\nu\alpha}\partial_\alpha] + [A_{\nu\gamma}\partial^\gamma, G^{\mu\nu\alpha}\partial_\alpha] \\
&= i(\partial_\nu F^{\mu\nu} + \partial_\nu G^{\mu\nu} + F^{\mu\nu\beta}\partial_\beta A_\nu + G^{\mu\nu\beta}\partial_\beta A_\nu + A_{\nu\gamma}\partial^\gamma F^{\mu\nu} + A_{\nu\gamma}\partial^\gamma G^{\mu\nu}) \\
&\quad + (\partial_\nu F^{\mu\nu\alpha} + \partial_\nu G^{\mu\nu\alpha} + A_{\nu\gamma}\partial^\gamma F^{\mu\nu\alpha} - F^{\mu\nu\gamma}\partial_\gamma A_\nu^\alpha \\
&\quad + A_{\nu\gamma}\partial^\gamma G^{\mu\nu\alpha} - G^{\mu\nu\gamma}\partial_\gamma A_\nu^\alpha)\partial_\alpha,
\end{aligned}
$$

$$(2.69)$$

so that the field equations become

$$
\partial_\nu F^{\mu\nu} + \partial_\nu G^{\mu\nu} + F^{\mu\nu\beta}\partial_\beta A_\nu + G^{\mu\nu\beta}\partial_\beta A_\nu + A_{\nu\gamma}\partial^\gamma F^{\mu\nu} + A_{\nu\gamma}\partial^\gamma G^{\mu\nu} = 0,
$$
$$
\partial_\nu F^{\mu\nu\alpha} + \partial_\nu G^{\mu\nu\alpha} + A_{\nu\gamma}\partial^\gamma F^{\mu\nu\alpha} - F^{\mu\nu\gamma}\partial_\gamma A_\nu^\alpha + A_{\nu\gamma}\partial^\gamma G^{\mu\nu\alpha} - G^{\mu\nu\gamma}\partial_\gamma A_\nu^\alpha = 0
$$

$$(2.70)$$

The first of the above equations describes the propagation of the electromagnetic field in the presence of gravity. It is immediately obvious that this equation is much more complicated than the propagation of the electromagnetic field in vacuum which is described by $\partial_\nu F^{\mu\nu} = 0$. The gravitational field obviously influences the electromagnetic field, just like in GR. However, the second of the two equations does not contain any electromagnetic field. That means that the propagation of the gravitational field is not affected at all by the presence of the electromagnetic field. This is a stark contrast with GR, where all matter produces a gravitational field, including the electromagnetic radiation.

The GR equation that governs propagation of the electromagnetic field in the presence of gravity is given by

$$
\partial_\nu F^{\mu\nu}_{GR} + \Gamma^\alpha_{\nu\alpha} F^{\mu\nu}_{GR} = 0, \tag{2.71}
$$

where all index contractions are done with the GR metric $g^{\mu\nu}(x)$, and $F^{\mu\nu}_{GR}$ is given as

$$F_{GR}^{\mu\nu} = \partial^{\mu}A^{\nu} - \partial^{\nu}A^{\mu} + \Gamma_{\alpha}^{\mu\nu}A^{\alpha} - \Gamma_{\beta}^{\nu\mu}A^{\beta}. \tag{2.72}$$

Thus, the field equations for the electromagnetic field A^{μ} are

$$\partial_{\nu}(\partial^{\mu}A^{\nu} - \partial^{\nu}A^{\mu} + \Gamma_{\alpha}^{\mu\nu}A^{\alpha} - \Gamma_{\beta}^{\nu\mu}A^{\beta}) + \Gamma_{\nu\gamma}^{\gamma}(\partial^{\mu}A^{\nu} - \partial^{\nu}A^{\mu} + \Gamma_{\alpha}^{\mu\nu}A^{\alpha} - \Gamma_{\beta}^{\nu\mu}A^{\beta}) = 0. \tag{2.73}$$

This is also a non-linear partial differential equation which reduces to Maxwell equations in the low-energy limit. However, due to the contraction of indexes through the GR metric tensor field, this equation is third order in gravitational field, not second like in the NL model above. (Christofell symbols are linear in gravitational field.) These higher order terms would only be observable in very strongly interacting systems - mergers of two rapidly spinning black holes, for instance. At present there does not exist even a numerical method that is robust enough to tackle these kinds of problems, and thus, it would be impossible to compare the predictions of the two theories. In the framework of linearized gravity, though, both equations predict that the characteristics of the electromagnetic field equations are described by the geodesics.

The field equations for the gravitational field in GR are given by the Einstein field equations ([24, 23]):

$$R^{\mu\nu} - \frac{1}{2}g^{\mu\nu}R = 8\pi T^{\mu\nu}, \tag{2.74}$$

where $R_{\mu\nu}$ is the Ricci curvature tensor, R is the curvature scalar, and $T^{\mu\nu}$ is the stress-energy tensor of the electromagnetic field given by

$$T^{\mu\nu} = F^{\mu\alpha}F_{\alpha}^{\nu} - \frac{1}{4}F_{\alpha\beta}F^{\alpha\beta}g^{\mu\nu}. \tag{2.75}$$

The stress-energy tensor acts as a source of the gravitational field. However, this has only been observed directly in the case of matter, not for the radiational stress-energy. The amounts of pure radiation that would observably alter the gravitational field in its vicinity has probably not been seen since the Big Bang.

25

2.4.4 The Field Equations in the Presence of the Matter Field

We have postponed the treatment of the interaction of of the non-local fields with matter for two reasons. First, we wanted to see what is the meaning of having a field equations for the derivative-valued fields. These field equations can be directly reduced to the partial differential equations of the ordinary local higher spin fields. The second reason is a bit more technical: the strighforward symbolic manipulation of the action that yielded the gauge field equations seems to imply that the source for the gauge fields is a bi-local quantity, which is not easily relatable to the differential operators.

In order to overcome the second dificulty, we will assume that the action for the non-local gauge fields (2.13) is a functional of the local gauge filds ($A^\mu(x)$ and $A^{\mu\nu}(x)$ in our case), and that the extremization of this action in the absence of matter fields yields the usual equations of motion (2.70).

The matter fields interact with gauge fields throu a covariant derivative. For instance, the gravitational interaction comes from the term

$$\int \bar\psi(x)\gamma^\mu A_{\mu\nu}(x)\partial^\nu\psi(x)d^dx.$$

Varying this term with respect to $A_{\mu\nu}(x)$ gives us the following quantity:

$$P^{\mu\nu} = \bar\psi(x)\gamma^\mu\partial^\nu\psi(x),$$

so that the interaction between matter and gravity is given by

$$\partial_\mu F^{\mu\nu\alpha}+\partial_\mu G^{\mu\nu\alpha}+A_{\mu\gamma}\partial^\gamma F^{\mu\nu\alpha}-F^{\mu\nu\gamma}\partial_\gamma A_\mu^\alpha+A_{\mu\gamma}\partial^\gamma G^{\mu\nu\alpha}-G^{\mu\nu\gamma}\partial_\gamma A_\mu^\alpha = \bar\psi(x)\gamma^\mu\partial^\nu\psi(x).$$

$$(2.76)$$

We see that $P^{\mu\nu}$ acts as a source of the gravitational field. Its role is comparable to that of the stress-energy tensor in GR. In fact, $P^{\mu\nu}$ is mathematically equivalent to a part of the stress-energy tensor of for Dirac field, which is given by

$$T^{\mu\nu} = \bar\psi(x)\gamma^\mu\partial^\nu\psi(x) + \bar\psi(x)\gamma^\nu\partial^\mu\psi(x) - m\bar\psi(x)\psi(x)\eta^{\mu\nu}$$

In the non-relativistic limit only the T^{00} component of the stress-energy tensor will have any effect on the gravitational field. In that limit, for sta-

26

tionary point-like sources, both GR and the non-local gauge field theory will have an equivalent form of the source term, given by

$$P^{00} \approx T^{00} \approx m\delta(x - y). \tag{2.77}$$

This is another confirmation of the agreement of non-local gauge field theory with GR in this limit.

2.4.5 Putting Everything Together

The above section has shown us some major parallels between the non-local gravitational field and the electromagnetic field. It is tempting to combine those two fields into a single entity. They are both hermitian, spin-one fields. In fact, the most general non-local gauge field is a spin-one hermitian operator.

If we were to put the field on a lattice, the non-local gauge field would be represented by an infinite dimensional matrix. The electromagnetic field would then be a matrix with only diagonal elements, and the gravitational field would involve the next-to-diagonal matrix elements such that the entire matrix is hermitian. The most general hermitian matrix would involve elements arbitrarily far from the diagonal, and in the continuum limit these elements would correspond to the derivative terms at all orders.

This leads us to adopt the following definition of the non-local gauge field operator:

$$\hat{A}^\mu = \sum_{k=0}^{\infty} A^\mu_{(k)}(x)\mathcal{D}^{(k)}, \tag{2.78}$$

where

$$A^\mu_{(0)}(x) = A^\mu,$$
$$A^\mu_{(k)}(x) = A^\mu_{\nu_1\nu_2...\nu_k},$$
$$\mathcal{D}^{(0)} = \hat{\mathbb{1}},$$
$$\mathcal{D}^{(k)} = (-i\partial^{\nu_1})(-i\partial^{\nu_2})\dots(-i\partial^{(k)}),$$

with the additional hermiticity requirement

$$\partial^{\nu_j} A^{\mu}_{\nu_1 \ldots \nu_j \ldots \nu_k} = 0$$

for all ν_j.

The derivative terms have the following interpretation: the higher the derivative term of the operator, the less local is its action. The electromagnetic interaction has no derivative terms and therefore it corresponds to the local interaction. To simplify notation throughout this thesis, we have used dimensionless units. However to appreciate the increasing smallness of the non-local interactions, we just need to multiply each derivative with a constant on the order of the Planck length. Therefore, the additional interaction terms are increasingly smaller by enormous orders of magnitude and these interactions would be impossible to observe at the ordinary energy scales. Thus, the Non-Local Unitary Gauge Theory when interpreted in this way is a legitimate candidate for a real physical theory.

2.5   Interpretation of Non-Local Operators

At this point, the non-local operators that we have introduced are primarily motivated by the underlaying mathematical analogy with the local gauge theories. Furthermore, it may not be clear how one is to think of the derivative-valued classical fields. It would serve a pedagogical purpose to try to make some additional descriptions and clarifications of the non-local field. In particular, it is necessary to elaborate a bit further on their relation to the global and local transformations.

The rationale for my line of research can be summed up in one simple principle: treat fields as elements of a vector space. In the case of fields over a continuous spacetime, this means that continuous coordinates will play an analogous role that indices play for finite-dimensional vectors.

The interpretation of the derivative-valued transformations is not as straightforward. If our fields were finite-component discrete vectors, then the most natural way of expanding an arbitrary matrix transformation would be to write it in terms of a basis of matrices. For instance, an arbitrary 2×2 Hermitian matrix can be expanded in terms of Dirac σ matrices, and similarly for any $su(N)$ algebra. In the case of infinite discrete matrices this procedure

28

becomes unwieldy, and it is not clear that for integral or differential operators it would even be possible to define the corresponding basis operators.

As it was shown before, the procedure that is adopted in this work is to expand the non-local operators as a series of higher and higher derivative operators. We interpret the higher derivative operators roughly as less and less local. The motivation for this is the scheme for writing the approximation for derivative that is used in the numerical treatment of differential equations. For instance, the first derivative of a field on a lattice can be expressed as $v_n \rightarrow v_{n+1} - v_n$. In terms of matrices, this will look something like

$$
\begin{pmatrix} v_1 \\ v_2 \\ v_3 \\ v_4 \\ \vdots \end{pmatrix} \rightarrow \begin{pmatrix} -1 & 0 & 0 & 0 & \dots \\ 1 & -1 & 0 & 0 & \dots \\ 0 & 1 & -1 & 0 & \dots \\ 0 & 0 & 1 & -1 & \dots \\ \vdots & \vdots & \vdots & \vdots & \ddots \end{pmatrix} \begin{pmatrix} v_1 \\ v_2 \\ v_3 \\ v_4 \\ \vdots \end{pmatrix}
$$

We see that the above representation of the derivative includes diagonal and first off-diagonal elements. In terms of the conventions used in this work, the derivative operator is thus clearly a non-local operator, at least in position space representation. This is at odds with the prevailing convention in the physics literature, where differential operators are considered local, and the term non-local is exclusively reserved for integral operators. The distinction between differential and integral operators does have a basis in mathematics, and there are non-trivial considerations that need to be taken into the account when dealing with them. However, from a purely conceptual standpoint, differential and integral operators have much more in common than either of them has with local operators as defined here.

It is easy to see now how a general differential operator would roughly correspond to a matrix with arbitrary diagonal and first off-diagonal elements. Similarly, the second order differential operator will correspond to a matrix with non-zero first and second off-diagonal elements, and similarly for higher derivative operators. We can write the corresponding matrices as follows.

Local:

$$\begin{pmatrix} a_{11} & 0 & 0 & 0 & \dots \\ 0 & a_{22} & 0 & 0 & \dots \\ 0 & 0 & a_{33} & 0 & \dots \\ 0 & 0 & 0 & a_{44} & \dots \\ \vdots & \vdots & \vdots & \vdots & \ddots \end{pmatrix} \sim A^{\mu}(x)$$

One derivative:

$$\begin{pmatrix} a_{11} & a_{12} & 0 & 0 & \dots \\ a_{21} & a_{22} & a_{23} & 0 & \dots \\ 0 & a_{32} & a_{33} & a_{34} & \dots \\ 0 & 0 & a_{43} & a_{44} & \dots \\ \vdots & \vdots & \vdots & \vdots & \ddots \end{pmatrix} \sim i A^{\mu\nu}(x)\partial_{\nu}$$

Two derivatives:

$$\begin{pmatrix} a_{11} & a_{12} & a_{13} & 0 & \dots \\ a_{21} & a_{22} & a_{23} & a_{24} & \dots \\ a_{31} & a_{32} & a_{33} & a_{34} & \dots \\ 0 & a_{42} & a_{43} & a_{44} & \dots \\ \vdots & \vdots & \vdots & \vdots & \ddots \end{pmatrix} \sim -A^{\mu\nu\rho}(x)\partial_{\nu}\partial_{\rho}$$

From this consideration it is easy to see how the most general infinite-dimensional matrix would correspond to a differential operator of an infinitely high order. This is precisely the expansion of the Hermition operators in terms of higher and higher differential operators that we have introduced in this chapter.

2.6 Other examples of non-local unitary transformations used in Physics

Non-local transformations are fairly common in physics, although they are usually not explicitly viewed as such. In our discussion of non-local gauge theories we dealt with diffeomorphism invariance as one class of such transformations, and this led to our formulation of non-local unitary gauge theory. However, there are a few other non-local linear transformations that are even more prevalent, and it is our intention in this section to make their relation to non-local gauge transformations more explicit. The following two subsections deal with non-local transformations that need to be represented in terms of an integral kernel, so they differ from the differential operator expansion that was introduced in the previous section. Nevertheless, they too are unitary transformations that are relevant to physics.

2.6.1 Fourier transforms

Fourier transformations are usually thought of as a computational device that is useful for diagonalizing a differential operator. In quantum mechanics Fourier transforms are used to go back and forth from position-space representation to momentum-space transformation. Other than that, they are usually not thought of as having any deep physical meaning comparable to the transformations associated with other symmetry groups.

In this section we will explicitly treat Fourier transforms as non-local unitary transformations. This will make it clear that they are just a special subgroup of the most general non-local unitary group.

The integral kernel of the Fourier transforms can be written as

$$U(x, y) = (2\pi)^{-D/2} e^{ixy},$$

and its Hermitian conjugate is

$$U^*(y, x) = (2\pi)^{-D/2} e^{-iyx},$$

so that

$$\int U(x, z) U^*(z, y) dz = (2\pi)^{-D} \int e^{ixz} e^{-izy} dz$$
$$= (2\pi)^{-D} \int e^{iz(x-y)} dz \qquad (2.79)$$
$$= \delta(x - y),$$

where we have recognized the integral representation of the Dirac delta function to get to the last line. This simple exercise shows that Fourier transforms are unitary non-local operators.

The re-writing of the Fourier transforms as a unitary gauge transforms has an important theoretical implication. In wave mechanics it is customary to talk about the equivalence of the position and momentum representations. However, once we move to QFT and the local gauge theories all quantities and transformations are expressed in terms of the position space. When we take into account the fact that Fourier transforms are just a particular kind of non-local unitary transforms, and that action should be invariant under the entire non-local unitary gauge group, then this apparent discrepancy is easily explained. From the point of view of non-local gauge theory, the local representation corresponds to a particular choice of gauge in which we work. Thus, although the particular form of action doesn't seem explicitly equivalent to the momentum space representation, the gauge invariance guarantees that the total action will be invariant.

2.6.2 Parity/Time reversal operators

Parity and Time reversal transformations rely on reversal of the sign of coordinates of the field configuration: $\phi(x) \rightarrow \phi(-x)$. The kernel of this operator is just

$$U(x, y) = \delta(x + y) = (2\pi)^{-D} \int e^{ik(x+y)} dk,$$

and its Hermitian conjugate is

$$U^*(y, x) = (2\pi)^{-D} \int e^{-ik(y+x)} dk = \delta(y + x),$$

so that

$$U(x, z)U^*(z, y) = \int \delta(x + z)\delta(z + y) dz = \delta(x - y), \qquad (2.80)$$

which proves the fact that this is a unitary transformation.

From the above consideration it is straightforward to conclude that $\delta(x - y)$ and $\delta(x + y)$ form a two-element discrete group. It can also be shown that these transformations belong to a continuous one parameter subgroup of general non-local unitary transformations.

Let us take a look at the following kernel

$$U(x, y) = \delta(x - e^{i\theta}y) = \delta(x - \cos\theta y - i\sin\theta y),$$

32

where $\theta \in \mathbb{R}$. We see that for $\theta = 2n\pi$, $U(x,y) = \delta(x-y)$, and for $\theta = (2n+1)\pi$, $U(x,y) = \delta(x+y)$, where $n \in \mathbb{N}$. It remains to prove that $U(x,y)$ is a kernel of a unitary operator.

We have

$$U^*(y,x) = (2\pi)^{-D} \int e^{-ik(e^{-i\theta}y-x)}dk = \delta(e^{-i\theta}y - x),$$

so that

$$U(x,z)U^*(z,y) = \int \delta(x - e^{i\theta}z)\delta(z - e^{-i\theta}y)dz = \delta(x-y). \qquad (2.81)$$

Since parity/time reversal symmetry is just a special case of a non-local unitary group, it follows that any theory that is gauge-invariant under the more general group will necessarily be invariant under parity/time reversal. However, if the gauge symmetry is somehow broken, then the ground state propagator may not be explicitly invariant under the parity/time reversal symmetry. We see that the considerations of CPT symmetry are closely linked with the overall gauge group of the theory.

2.6.3 Poincaré group

The Poincaré group is usually thought of as an ordinary ten-parameter Lie group (four translations and six rotations and boosts). However, it is equally possible to think of it as a subgroup of "local translatons", that is, the group of unitary transformations generated by a hermitian operator of the form

$$\theta^\mu(x)\partial_\mu,$$

with $\partial_\mu\theta^\mu(x) = 0$. In the case of the Poincaré group we have

$$\theta^\mu(x) = p^\mu + \omega^{\mu\nu}x_\nu, \qquad (2.82)$$

with $\omega^{\mu\nu} = -\omega^{\nu\mu}$.

In the discussion of non-local fields one common objection to the idea of using them as physical entities is the supposition that they lead to a loss of causality. This, however, will not be the case on a classical level if we assume that they transform in the adjoint representation of the Poincaré group.

The most common way of defining the adjoint representation in physics is through the use of group structure constants. However, this is not easily achieved for groups with continuously infinite number of generators. Under those circumstances the more appropriate definition of the adjoint representation would be through the use of the commutator transformation. That is, if $\hat{\theta}$ is a generator of the group in the fundamental representation, then $[\hat{\theta}, \ ]$ is the generator of the adjoint representation. While $\hat{\theta}$ is a matrix (resp. operator) acting on vectors (functions), $[\hat{\theta}, \ ]$ is a commutator acting on matrices (operators).

Adjoint representation of the Poincaré group. .

If an operator $\hat{\phi} = \phi^{\mu_1 \cdots \mu_n}(x)\partial_{\mu_1} \ldots \partial_{\mu_n}$ transforms as a scalar in the adjoint representation of the Poincaré group, then the tensor field $\phi^{\mu_1 \cdots \mu_n}(x)$ transforms as a tensor of rank n in the fundamental representation of the group.

Proof. We will first show it on the example of fields of the form $\phi^\mu(x)\partial_\mu$. (These would be the standard Lorentzian vector fields.) We will show it for the case of the Lorentz transformations, since they are the only ones that involve spin degrees of freedom.

The adjoint action of the transformation $\omega_{\mu\nu}x^\mu\partial^\nu$ on the field $\phi^\mu(x)\partial_\mu$ is given by:

$$
\begin{aligned}
[\omega^{\mu\nu}x^\mu\partial^\nu, \phi^\alpha\partial_\alpha] &= \omega_{\mu\nu}(x^\mu\partial^\nu\phi^\alpha\partial_\alpha + x^\mu\phi^\alpha\partial^\nu\partial_\alpha - \phi^\alpha\partial_\alpha x^\mu\partial^\nu - x^\mu\phi^\alpha\partial^\nu\partial_\alpha) \\
&= \omega_{\mu\nu}(x^\mu\partial^\nu\phi_\alpha\partial^\alpha - \phi^\mu\partial^\nu) \\
&= \omega_{\mu\nu}(x^\mu\partial^\nu\phi_\alpha - \phi^\mu\delta^\nu_\alpha)\partial^\alpha,
\end{aligned}
$$

(2.83)

from where we see that $\phi^\mu(x)$ transforms as

$$
\phi^\mu(x) \rightarrow \omega_{\alpha\beta}(x^\alpha\partial^\beta\phi^\mu(x) - S^{\alpha\beta\mu\nu}\phi_\nu(x)),
$$

(2.84)

where

$$
S^{\alpha\beta\mu\nu} = \eta^{\alpha\mu}\eta^{\beta\nu} - \eta^{\alpha\nu}\eta^{\beta\mu}.
$$

Now we will take a look at the most general case. We have

$$[\omega^{\mu\nu}x^\mu\partial^\nu, \phi^{\alpha_1\ldots\alpha_n}\partial_{\alpha_1}\ldots\partial\alpha_n] = \omega_{\mu\nu}(x^\mu\partial^\nu\phi^{\alpha_1\ldots\alpha_n} - \phi^{\mu\alpha_2\ldots\alpha_n}\eta^{\nu\alpha_1}$$

$$- \phi^{\mu\alpha_2\ldots\alpha_n}\eta^{\nu\alpha_2}$$

$$- \ldots$$

$$- \phi^{\alpha_1\ldots\alpha_{k-1}\mu\alpha_{k+1}\ldots\alpha_n}\eta^{\nu\alpha_k}$$

$$- \ldots$$

$$- \phi^{\alpha_1\ldots\alpha_{n-1}\mu}\eta^{\nu\alpha_n})\partial_{\alpha_1}\ldots\partial_{\alpha_n},$$

$$(2.85)$$

from where we see that $\phi^{\mu_1\ldots\mu_n}(x)$ transforms as

$$\phi^{\mu_1\ldots\mu_n}(x) \rightarrow \omega_{\alpha\beta}(x^\alpha\partial^\beta\phi^{\mu_1\ldots\mu_n}(x) - S^{\alpha\beta\mu_1\nu_1\ldots\mu_n\nu_n}\phi_{\nu_1\ldots\nu}(x)), \qquad (2.86)$$

where

$$S^{\alpha\beta\mu_1\nu_1\ldots\mu_n\nu_n} = (\eta^{\alpha\mu_1}\eta^{\beta\nu_1} - \eta^{\alpha\nu_1}\eta^{\beta\mu_1})\eta^{\mu_2\nu_2}\ldots\eta^{\mu_n\nu_n}$$

$$+ \eta^{\mu_1\nu_1}(\eta^{\alpha\mu_2}\eta^{\beta\nu_2} - \eta^{\alpha\nu_2}\eta^{\beta\mu_2})\eta^{\mu_3\nu_3}\ldots\eta^{\mu_n\nu_n}$$

$$+ \ldots$$

$$+ \eta^{\mu_1\nu_1}\ldots\eta^{\mu_{n-1}\nu_{n-1}}(\eta^{\alpha\mu_n}\eta^{\beta\nu_n} - \eta^{\alpha\nu_n}\eta^{\beta\mu_n})$$

$$= \sum_{k=1}^{n}\eta^{\mu_1\nu_1}\ldots\eta^{\mu_{k-1}\nu-1}(\eta^{\alpha\mu_k}\eta^{\beta\nu_k} - \eta^{\alpha\nu_k}\eta^{\beta\mu_k})\eta^{\mu_{k+1}\nu_{k+1}}\ldots\eta^{\mu_n\nu_n}.$$

$$(2.87)$$

$S^{\alpha\beta\mu_1\nu_1\ldots\mu_n\nu_n}$ is the spin tensor operator for the spin-n field, and thus we prove the theorem.

2.6.4 Conformal Group

The conformal group is an extension of the Poincaré group whose Lie algebra in addition to operators P^μ and $M^{\mu\nu}$ also has operators D (dilatation) and K^μ (special conformal transformation). In the functional space representation, D and K^μ are also first order differential operators:

$$D = -ix^\mu \partial_\mu,$$

$$K_\mu = -i(2x_\mu x^\nu \partial_\nu - x^2 \partial_\mu) = -i(2x_\mu x^\nu - x^2 \delta^\nu_\mu)\partial_{nu},$$

(2.88)

and thus we would expect the Conformal group to also be a subgroup of "local translations" generated by the $g^\mu(x)\partial_\mu$. However, unlike the case of the Poincaré group, these additional generators are not Hermitian in the most general case and thus this group is not unitary. This follows from the fact that in general $\partial_\mu g^\mu(x) \neq 0$ for dilatations. For dilatations we have $g^\mu(x) = x^\mu$ and therefore $\partial_\mu g^\mu = d$, which implies that dilatations are not unitary in any spacetime dimensions. On the other hand, for special conformal transformation we have:

$$\partial_\nu(2x^\mu x^\nu - x^2 \eta^{\mu\nu}) = 2x^\mu d - 2x^\mu = 2x^\mu(d-1),$$

(2.89)

which makes this operator Hermitian for $d = 1$.

From the above discussions it is easy to see why only special forms of action will be invariant under the conformal symmetry. In our treatment of the non-local gauge field theory we treated the mass term as the inner product term for fields, and thus we sought to construct field theories that will keep this term invariant. In general, conformal invariance is associated with massless fields, and thus there is no requirement for the invariance group to be unitary.

3 Effective field interactions and the non-local interaction term

3.1 Introduction

Quantization of the gravitational force is one of the outstanding problems in theoretical physics [36, 37, 38, 39, 40]. One of the main conceptual obstacles to the program of perturbative quantization of gravity is the existence of the dimensionful interaction constant (Newton's constant). Because of this feature of gravity, higher and higher terms in the perturbative expansion require the introduction of more and more counterterms in the original Lagrangian, thus rendering the whole procedure meaningless.

The theory of weak interactions also started out as a theory with a dimensionful interaction constant - Fermi's constant. It was originally formulated as what would be called today "four-Fermi theory" [42, 43]. With the advent of quantum field theory (QFT) this theory was also shown to be mathematically inconsistent, and the search began for an alternative formulation of the theory of weak interactions. The theory that was eventually developed was based on massive intermediate gauge bosons [44, 45, 46]. The mass of these bosons is related to Fermi's constant, and the perturbation parameter is now a new dimensionless constant.

It would be ideal if we could do something similar with the gravitational interaction. At first sight this seems impossible: we know that the gravitational interaction has an infinite range, so massive gravity would not solve this problem. Furhermore, the effective gravitational interaction term contains derivatives, and it is unclear how to deal with them. To shed some light on these problems, I will present the effective interaction terms for various classical field theories. By carefully investigating the mathematical form of this term in the momentum space, I will attempt to show how it leads to a general four-point function for theories with derivative interactions. Thus, if we were to look for an intermediate boson reformulation of this theory, we would need to consider non-local (bi-local) fields. Following this, I argue that by combining the non-local with the local field it is possible to come up with a theory with a dimensionless interaction constant.

Some of the mathematical considerations in the first part of this paper are very straightforward, almost pedestrian. However, in the interest of physics (as opposed to mathematics), it is important to go through them. This is because many of the physical terms used in this article have a somewhat different meaning than the one in the rest of the physics literature, rendering it a useful pedagogical exercise to re-introduce them from scratch. For instance, a differential operator is usually thought of as a local operator, and by rewriting it in integral form I show explicitly how it differs from non-derivative local operators.

3.2 Momentum-space form of the interaction term

I would like to shed some more light on the difference between local and derivative interaction terms. In order to achieve this it is appropriate to treat both local and derivative interactions within the same mathematical framework. I will demonstrate that this can be achieved if we work in the momentum-space representation. In this way, the mathematical differences between those two theories will become transparent. To show this clearly, it is necessary to Fourier transform the interaction term and to maintain the final momentum-space form without any simplifications and reinterpretations. We will sacrifice computational brevity for the sake of subsequent mathematical transparency.

3.2.1 Local Interaction

We start with a local interaction and show that in momentum space it can be written as a four-point functor, albeit of a very special kind. The special nature of this functor allows us to treat it as a function of single momentum, and thus to reinterpret it as a Green's function of an appropriate differential operator.

Scalar Field

The usual form of the interaction term in any local field theory reads as

$$H_{\text{int}} = \int dxdy J(x)U(x-y)J(y), \tag{3.1}$$

where we have implicitly assumed that all integrals are over D-dimensional spacetime.

For scalar field theory $J(x)$ is just a scalar current given by $J(x) = \psi^*(x)\psi(x)$, so that the above equation can be written as

$$H_{\text{int}} = \int dxdy \psi^*(x)\psi(x)U(x-y)\psi^*(y)\psi(y). \tag{3.2}$$

We would like to write the above term in momentum space and explicitly take into consideration the fact that this is a four-point interaction. To begin, we re-write all the fields from the above equation in terms of their Fourier transforms. We have

$$\psi(x) = (2\pi)^{-D/2}\int \hat{\psi}(k)e^{ixk}dk,$$

$$\psi^*(x) = (2\pi)^{-D/2}\int \hat{\psi}^*(k)e^{-ixk}dk, \tag{3.3}$$

$$U(x-y) = (2\pi)^{-D/2}\int \hat{U}(k)e^{ik(x-y)}dk.$$

We can therefore re-write equation (3.1) as

$$H_{\text{int}} = (2\pi)^{-5D/2}\int dxdydk_1dk_2dl_1dl_2dp\,\hat{\psi}^*(k_1)\hat{\psi}(k_2)\hat{U}(p)\hat{\psi}^*(l_1)\hat{\psi}(l_2)e^{-i[x(k_1-k_2-p)+y(l_1-l_2+p)]}$$

$$= (2\pi)^{-3D/2}\int dk_1dk_2dl_1dl_2\,\hat{\psi}^*(k_1)\hat{\psi}(k_2)\hat{U}(l_2-l_1)\delta(l_1-l_2+k_1-k_2)\hat{\psi}^*(l_1)\hat{\psi}(l_2), \tag{3.4}$$

and we see that the four-point interaction term has the following form:

$$\begin{aligned}U_{\text{int}}(k_1,k_2,l_1,l_2) &= (2\pi)^{-3D/2}\hat{U}(k_1-k_2)\delta(l_1-l_2+k_1-k_2)\\&\equiv (2\pi)^{-3D/2}\hat{U}(l_2-l_1)\delta(l_1-l_2+k_1-k_2).\end{aligned} \tag{3.5}$$

It is clear that the above interaction term is a four-point functor. However, the delta function is just assumed to accompany any interaction that conserves momentum, and therefore it is usually left out of the description. We are therefore left with $\hat{U}(p)$, which is a regular function of single momentum. This function is usually associated with a kernel of a Green's function of an appropriate intermediate interaction field.

A particular form of interaction that we encounter all the time in physics is the inverse-square potential:

$$U(x-y) = \frac{e^2}{(x-y)^2} \Rightarrow U_{\text{int}}(k_1, k_2, l_1, l_2) = \frac{1}{(2\pi)^{3D/2}} \frac{e^2}{(k_1 - k_2)^2 + i\epsilon} \delta(l_1-l_2+k_1-k_2).$$

(3.6)

This interaction term corresponds to a Green's function of d'Alembertian differential operator.

Dirac Field

The case of Dirac field is a bit more complicated due to the presence of spin degrees of freedom. In the most general case, the local field interaction should be represented in terms of rank-four four-point tensor $U_{abcd}(k_1, k_2, l_1, l_2)$, where indexes a, b, c, d correspond to spin degrees of freedom. The interaction term looks like

$$H_{\text{int}} = \frac{1}{(2\pi)^{3D/2}} \int dk_1 dk_2 dl_1 dl_2 \hat{\psi}_a^*(k_1)\hat{\psi}_b(k_2)\hat{U}^{abcd}(k_1, k_2, l_1, l_2)\hat{\psi}_c^*(l_1)\hat{\psi}_d(l_2).$$

(3.7)

In the Standard Model, however, it is customary to write the source term of the interaction in terms of Dirac gamma matrices (γ^μ) and their Clifford algebra. In the case of QED we have a vector-vector interaction term

$$H_{\text{int}} = \int dx dy J^\mu(x) U_{\mu\nu}(x - y) J^\nu(y),$$

(3.8)

where $J^\mu(x) = \bar{\psi}^a(x)\gamma^\mu_{ab}\psi^b(x)$, and $\bar{\psi}_a = \psi^{*b}\gamma^0_{ba}$. Thus, instead of spin indexes, all interaction terms are written in terms of vector indexes. For QED the effective interaction term has the form

$$U_{\mu\nu}(k_1, k_2, l_1, l_2) = \frac{1}{(2\pi)^{3/2}} \frac{e^2 \eta_{\mu\nu}}{(k_1 - k_2)^2 + i\epsilon} \delta(l_1 - l_2 + k_1 - k_2), \qquad (3.9)$$

where the result is written in Lorentz gauge, and $\eta^{\mu\nu}$ is the flat metric tensor.

3.2.2 Non-Local (Derivative) Interaction

It is fairly straightforward to modify the interaction term to include derivative intercations. For instance, the derivative-derivative interaction term can be written as

$$H'_{int} = \int dx dy \psi^*(x) \partial_x \psi(x) U(x - y) \psi^*(y) \partial_y \psi(y). \qquad (3.10)$$

In momentum representation, the introduction of the derivative terms will manifest itself as extra factors of incoming and/or outgoing momenta. For instance, one can see by inspection that H'_{int} will have the following form in momentum space:

$$H'_{int} = -\frac{1}{(2\pi)^{3/2}} \int dk_1 dk_2 dl_1 dl_2 \hat{\psi}^*(k_1) \hat{\psi}(k_2) \hat{U}(l_2 - l_1) k_2 l_2 \delta(l_1 - l_2 + k_1 - k_2) \hat{\psi}^*(l_1) \hat{\psi}(l_2),$$
$$(3.11)$$

which leads to the following form of the four-point interaction term:

$$
\begin{aligned}
U'_{int}(k_1, k_2, l_1, l_2) &= -\frac{1}{(2\pi)^{3/2}} \hat{U}(k_1 - k_2) k_2 l_2 \delta(l_1 - l_2 + k_1 - k_2) \\
&\equiv -\frac{1}{(2\pi)^{3/2}} \hat{U}(l_2 - l_1) k_2 l_2 \delta(l_1 - l_2 + k_1 - k_2).
\end{aligned}
\qquad (3.12)
$$

It is clear that the above form of the interaction has a fundamentally different functional form than the local case. In the local case, the interaction potential only depended functionally on the momenta in the combination $l_1 - l_2$, or equivalently $k_1 - k_2$. This enabled us to relate the effective four-point interaction term to a propagator of a local gauge field that was given by a two-point function.

We see that in the case of derivative interactions, it will in general be impossible to directly identify the effective interaction term with a propagator of a local field. It is possible to remedy this by introducing a propagator for

41

a non-local (bi-local) field. Mathematically this propagator would need to be a four-point function, which is exactly what we have for the most general interaction.

At this point this line of reasoning seems very mathematical, and it is not clear that it would have any relevant physical meaning. In the next section we will address some of those concerns and try to show that we can understand the effective interaction of the General Relativity in terms of a non-local propagator.

3.2.3   Effective Interaction in General Relativity

In terms of the effective interaction, we can understand General Relativity (GR) in terms of a coupling between two stress-energy tensors. In other words, the source current for GR is the stress-energy tensor. In general, the stress-energy tensor is a quantity that is bilinear in source fields. In that respect it is analogous to the gauge currents in the Standard Model. However, the stress-energy tensor is always composed of derivative terms. It is also a spin-two quantity, or equivalently a rank-two tensor. The exact form of this tensor will depend on the exact nature of the interacting fields. In the following sections we will consider the cases of massless scalar and Dirac fields.

The fact that the stress-energy tensor is a spin-two quantity has implication for the form of the interaction term. The interaction between two rank-two tensors needs to be mediated through a rank four tensor. In GR this tensor is related to the propagator of the gravitational field, and because of the universal nature of this field it has the same form for all species of particles. In a Lorentz-like gauge, this propagator has the following form:

$$\hat{D}^{\mu\nu\rho\sigma}(p) = \frac{\eta^{\mu\nu}\eta^{\rho\sigma} + \eta^{\mu\rho}\eta^{\nu\sigma} + \eta^{\mu\sigma}\eta^{\rho\nu}}{p^2 + i\epsilon}. \tag{3.13}$$

In terms of this propagator we can write the interaction term of the effective action as

$$S_{\text{int}} = \int dx dy T^{\mu\nu}(x) D_{\mu\nu\rho\sigma}(x - y) T^{\rho\sigma}(y), \tag{3.14}$$

where $T^{\mu\nu}(x)$ is the stress-energy tensor of the particular interacting field. In the next two sections we will deal with two forms of this tensor.

Massless Scalar Field

For the sake of computational simplicity, we will work with the massless scalar field. The stress-energy tensor of the massless scalar field is given by

$$T^{\mu\nu}(x) = \partial^\mu \phi(x)\partial^\nu \phi(x) - \frac{1}{2}\eta^{\mu\nu}\partial^\rho \phi(x)\partial_\rho \phi(x).\tag{3.15}$$

The interaction term in this case will have the form

$$
\begin{aligned}
S_{\text{int}} &= \int dk_1 dk_2 dl_1 dl_2 T^{\mu\nu}(k_1, k_2) D_{\mu\nu\rho\sigma}(k_1, k_2, l_1, l_2) T^{\rho\sigma}(l_1, l_2) \\
&= \int dk_1 dk_2 dl_1 dl_2 \phi(k_1)\phi(k_2) U_{int}(k_1, k_2, l_1, l_2), \phi(l_1)\phi(l_2)
\end{aligned}
\tag{3.16}
$$

where now

$$U_{\text{int}}(k_1, k_2, l_1, l_2) = \frac{k_1^\alpha l_{1\alpha} k_2^\beta l_{2\beta} + k_1^\alpha l_{2\alpha} k_2^\beta l_{1\beta} + k_1^\alpha k_{2\alpha} l_1^\beta l_{2\beta}}{(k_1 - k_2)^2 + i\epsilon}\delta(k_1 - k_2 + l_1 - l_2).\tag{3.17}$$

Thus, we see that we can re-write the effective GR interaction between two scalar fields as a non-local four-point interaction. This is just a particular instance of the general rule that we have pointed out earlier: we can re-write any local interaction between tensor currents made of derivative terms as a non-local interaction term. We do this by taking the derivative terms from the interaction sources and incorporating them into the interaction potential. The interaction term, on the other hand, is usually associated with another field that mediates the interaction, and the interaction potential is interpreted as a propagator of the interaction field. From the above considerations it seems natural to look for a two-point (bi-local) field that would be responsible for the gravitational interaction. Before we delve deeper into this topic, we will extend the above arguments, with some modifications, to the case of the Dirac field.

Dirac Field

The stress-energy tensor of the massless Dirac field is given by

$$T^{\mu\nu}(x) = \bar{\psi}(x)\gamma^\mu \partial^\nu \psi(x) + \bar{\psi}\gamma^\nu \partial^\mu \psi(x) - \eta^{\mu\nu}\bar{\psi}(x)\gamma^\alpha \partial_\alpha \psi(x).\tag{3.18}$$

43

As mentioned earlier in the Local Interaction section, the case of the Dirac field is slightly complicated due to presence of the spin indices. In principle, we could deal with these indices by making the interaction term a rank four tensor in spin space. However, in order to emphasize the link between the gravitational interaction and the QED interaction we will follow the same recipe as for the interaction potential and rewrite it as a Lorentz rank two tensor that mediates the interaction between two vector currents. Thus, we have

$$
\begin{aligned}
S_{\text{int}} &= \int dk_1 dk_2 dl_1 dl_2 T^{\mu\nu}(k_1, k_2) D_{\mu\nu\rho\sigma}(k_1, k_2, l_1, l_2) T^{\rho\sigma}(l_1, l_2) \\
&= \int dk_1 dk_2 dl_1 dl_2 \bar{\psi}(k_1)\gamma^\mu\psi(k_2) U_{\mu\nu}(k_1, k_2, l_1, l_2), \bar{\psi}(l_1)\gamma^\nu\psi(l_2),
\end{aligned}
\tag{3.19}
$$

where

$$
U^{\mu\nu} = \frac{k_2^\mu l_2^\nu + k_2^\nu l_2^\mu + \eta^{\mu\nu}k_2^\rho l_{2\rho}}{(k_1 - k_2)^2 + i\epsilon}\delta(k_1 - k_2 + l_1 - l_2).
\tag{3.20}
$$

When we compare equation (3.20) to equation (3.17), we see that the main difference is that in (3.20) we have terms that are proportional to a bilinear form $k_2^\mu l_2^\nu$, so unlike the scalar field case the interaction is quadratic, rather than quartic, in momenta. In the massive field case, these terms will manifest themselves as an interaction that is bilinear in masses of interacting particles in the low-energy regime. This is precisely the form of Newton's law of gravity. Thus, in the low-energy regime the most important feature of the gravitational interaction is its bilinear form.

In the discussion so far we have worked in dimensionless units. However, one of the main obstacles to the quantization of gravity thus far has been the fact that the interaction constant for gravitational interactions is dimension-ful, and thus the perturbative corrections to it introduce new terms with an arbitrarily high mass dimension. This is an undesirable property because it implies that we need to keep introducing higher and higher derivative terms into the original action. This property renders the theory unrenormalizable.

One of the main purposes of this article is to motivate a solution for this problem. The first part of the solution was the removal of the derivative terms from the field sources and their placement in the gauge-field propagator. The next step is the removal of the dimensionful interaction constant from interaction vertexes and its placement in the gauge field propagator.

44

The dimensionful constant that appears in the discussion of gravitational interaction is Newton's constant, G_N. In QFT this constant is related to Planck's mass, M_{pl}. We would like to motivate our discussion of the renormalizable gravitational interaction by pursuing the analogy with the Weak interaction.

3.2.4 Weak Interaction Analogy

Until the 1920s, the only two known physical interactions were the electromagnetic and gravitational interactions. The discovery of different types of radioactive decay introduced the notion of two more interactions. The so-called β decay was associated with the Weak nuclear interactions. One of the earliest theories of the Weak interaction was Fermi's four-fermion theory. It was a local field theory, and the interaction term for this can be written in the notation of this thesis as:

$$U_{\mu\nu}(k_1, k_2, l_1, l_2) = \frac{G_F}{\sqrt{2}}\eta^{\mu\nu}\delta(l_1 - l_2 + k_1 - k_2), \qquad (3.21)$$

where G_F is Fermi's constant. Thus we see that this theory was formulated in terms of current - current interactions. It was later realized that the exact form of the potential included an axial current as well. However, even before this realization, a far more serious problem with the theory was the fact that it did not lead to a consistent QFT. Fermi's constant is dimensionful, so the perturbation theory fails for the same reason as it does for quantum gravity: there is a need to introduce higher and higher interaction terms at each stage of the perturbative quantization expansion.

The way out of this problem was achieved by an introduction of additional Gauge fields that mediate the interaction. These gauge fields needed to be massive in order to explain their short range. In this view, the interaction term (3.21) was just an approximation for a more accurate effective interaction term

$$U_{\mu\nu}(k_1, k_2, l_1, l_2) = \frac{1}{(2\pi)^{3D/2}}\frac{e^2\eta_{\mu\nu}}{(k_1 - k_2)^2 + M^2_{Z^0,W^\pm} + i\epsilon}\delta(l_1 - l_2 + k_1 - k_2),$$

$$(3.22)$$

where now $M_{Z^0,W^\pm}$ denotes the mass of the Z^0 or $W^\pm$ gauge boson. In terms of the Fermi constant G_F we have

$$M_{Z^0,W^\pm} \propto e G_F^{-1/2}, \tag{3.23}$$

where the constant of proportionality depends on θ_W, the Weinberg angle. This interaction term is interpreted in terms of the propagator of intermediate vector bosons which has the following form

$$D^{\mu\nu} = \frac{\eta^{\mu\nu}}{p^2 + M_{Z^0,W^\pm}^2 + i\epsilon}. \tag{3.24}$$

Thus we see that we were able to take the dimensionful constant G_F from the interaction vertex and place it in the propagator of the gauge fields. This does not solve the whole problem of renormalizabuility of this theory (we still need to take advantage of the spontaneous symmetry breaking), but it does help with one important aspect. We have replaced a dimensionful expansion constant $G_F^{1/2}$, with a dimensionless one, e. This removes one important conceptual obstacle from constructing a perturbative expansion of this QFT.

We would like to apply the same strategy in order to construct a perturbative quantum gravity. A conceptually simple way of accomplishing this is to use the local interaction term and use the non-local part as a correction term. We can write

$$U_{\mu\nu}(k_1, k_2, l_1, l_2) = \frac{1}{(2\pi)^{3D/2}} \frac{e^2 \eta_{\mu\nu}}{(k_1 - k_2)^2 + i\epsilon} \frac{1}{1 + \frac{k_1 \cdot l_1}{e^2 M_{Pl}^2}} \delta(l_1 - l_2 + k_1 - k_2), \tag{3.25}$$

where M_{Pl} is Planck's mass.

The above interaction term has some interesting physical implications. We explicitly showed how it would be possible to put a dimensionful interaction constant M_{Pl} into what can be interpreted as a propagator of a non-local field. Furthermore, when we let $M_{Pl} \to \infty$ we recover a local field theory. Since Planck's mass is so much larger than other fundamental energy scales, this seems like a very desirable feature. However, the price that we had to pay for this was a much closer association between local and non-local interactions. That is, we need to view local and non-local interaction terms as part of the same interaction potential. In terms of the spin degrees of freedom, this interaction corresponds to an intermediate vector boson.

One of the peculiarities of this propagator is that it seems to correspond to a vector-vector interaction, even though it is well known that gravity is mediated through a tensor-tensor interaction. This apparent discrepancy can be easily resolved when we formulate the full formalism of the non-local gauge theory. In particular, we can show that the n-th order differential operator that transforms as a scalar in adjoint representation of the Poincaré group corresponds to rank-n tensor in the normal representation of the same group. Likewise, a vector field that is also a first order differential operator would correspond to a local tensor field.

If we expand the above propagator in terms of powers of $k_1 \cdot l_1$ we obtain gravity as a first order correction to the local propagator. The higher order terms will presumably correspond to higher-derivative corrections to the original interaction, and these terms will be additionally suppressed compared to gravity. At energy scales that are much lower than the Planck mass scale, these additional terms would not be directly observable. Their cumulative effect would however have an effect for interactions around and above the Planck energy scale, and due to the convergent nature of the whole sum it will have a non-divergent effect on the computation of the loop diagrams in the quantum theory. We will work this out in greater detail in the following chapters.

Another important observation concerning the above propagator is that we have made the coupling constant (e^2) part of the propagator. This ensures that the lowest order term in $k_1 \cdot l_1$ includes the local charge in its interaction (in our case QED), while the linear term does not include it. This will ensure that the gravitational interaction is not charge dependant. From the conceptual point however, this approach implies that gravity is a non-local correction to the local interaction, and presumably all local interactions have a non-local component. These interactions would not be observed because of the short range in the case of Weak interactions, and because of the confinement in the case of QCD. Future work will deal with these issues in greater depth.

The most natural way of introducing intermediate vector bosons is through the requirements of local gauge invariance and minimal coupling. However, for a non-local interaction local gauge invariance will not be sufficient; instead we need to impose a stronger requirement of non-local gauge invariance. This approach is developed in [21] as well as in the preceding chapter.

3.3   Form of integral operators that commute with the translation operator

3.3.1   Translationally invariant rank-2 operators

In this section we attempt to find the most general form of an integral operator that commutes with derivatives. That is, we look for the most general translationally invariant operator. Given an operator $\hat{U}$ that acts in a following manner on an arbitrary function $\psi(x)$

$$\psi(x) \rightarrow \int U(x,y)\psi(x), \tag{3.26}$$

we would like to have

$$[\partial, \hat{U}] = 0, \tag{3.27}$$

where ∂ is a differential operator in an arbitrary dimension. In order to systematically investigate this requirement, we express all operators in terms of their Fourier-transformed kernels.

Thus, we can write

$$\delta(x-y) = \int dk e^{-ik(x-y)} = \int dk dl \delta(k-l) e^{-i(kx-ly)}. \tag{3.28}$$

$$i\partial = \int dk k e^{-ik(x-y)}. \tag{3.29}$$

$$U(x,y) = \int dk dl e^{-i(kx-ly)} U^*(k,l). \tag{3.30}$$

We have

$$
\begin{aligned}
i\partial\hat{U} &= \int dk dl dm dz k e^{-ik(x-z)-i(lz-my)} U^*(l,m) \\
&= \int dk dl dm k \delta(k-l) e^{i(kx-my)} U^*(l,m) \\
&= -\int dl dm e^{-i(lx-my)} l U^*(l,m).
\end{aligned}
\tag{3.31}
$$

48

On the other hand

$$
\begin{aligned}
i\hat{U}\partial &= \int dk\,dl\,dm\,dz\, e^{-i(kx-lz)} U^*(k,l) m e^{-im(z-y)} \\
&= \int dk\,dl\,dm\, e^{-i(kx-my)} U^*(k,l) m \delta(l-m) \\
&= -\int dk\,dl\, e^{-i(kx-ly)} U^*(k,l) l \\
&= -\int dl\,dm\, e^{-i(lx-my)} U^*(l,m) m.
\end{aligned}
\tag{3.32}
$$

In order for equation (3.27) to be satisfied, we need to have

$$
l U^*(l,m) = U^*(l,m) m.
\tag{3.33}
$$

As a constraint on the function $U^*(l,m)$ the above equation cannot be satisfied. However, as a constraint on an integral kernel the above condition gives us

$$
U^*(l,m) = f(l)\delta(l-m) = f(m)\delta(l-m),
\tag{3.34}
$$

where $f(m)$ is an arbitrary function. If $f(m)$ is an analytic function, then the operator $\hat{U}$ turns out to be an arbitrary differential operator with constant coefficients.

Thus, if we are interested in building a translationally invariant action, we can make use of terms of the form

$$
\int dx\,dy\,\psi^*(x) U(x,y) \psi(y),
\tag{3.35}
$$

where $U(x,y)$ is given by equation (3.30). Clearly, free Dirac and Klein-Gordon fields satisfy this criterion.

3.3.2 Translationally invariant rank-4 operators

In order to construct a physically realistic action, we need to insert the interaction terms. In the simplest form, this is achieved by explicitly putting in a four-field term into the action. The simplest interacting theory is ϕ^4 theory, where the interaction term is proportional to a four-point delta function in position space.

49

It is also possible to model the Coulomb interaction as an interaction between two local currents. In that case, the four-field term reduces to two-point interaction term between two local currents. This has been explored in the earlier sections of this thesis. In fact, any bi-local interaction term that only explicitly depends on the difference between the two locations (i.e. has the form $U(x-y)$) will be translationally invariant.

In this section, we attempt to find the most general form of the four point interaction term that is translationally invariant. Our strategy will be the same as in the case of translationally invariant operators: we will write the interaction term in the momentum space, and then demand that it commutes with a particular form of differentiation operator.

We consider an interacting term of the following form:

$$\int dx\,dy\,dz\,dw\,\psi^*(x)\psi(y)U(x,y,z,w)\psi(z)\psi^*(w). \tag{3.36}$$

Again, we will take a look at the Fourier transform of the rank-4 operator $\hat{\mathbf{U}}$ which now has the form:

$$\hat{\mathbf{U}} \equiv \int dk\,dl\,dm\,dn\,U^*(k,l,m,n)e^{-i[kx-ly-mz+nw]} \tag{3.37}$$

Under the translation operator $\hat{T} = \exp(-ip \cdot \partial)$, $\hat{\mathbf{U}}$ transforms as

$$\hat{\mathbf{U}} \rightarrow \hat{T}^\dagger\hat{T}\hat{\mathbf{U}}\hat{T}\hat{T}^\dagger, \tag{3.38}$$

where the first two operators act on the first two coordinates of $\hat{\mathbf{U}}$, and the last two act on the third and fourth coordinate. We can also write the above equation in terms of exponentiation of commutators:

$$\hat{\mathbf{U}} \rightarrow \exp(-i[p \cdot \partial,\ ])\hat{\mathbf{U}}\exp(ip[\cdot\partial,\ ]), \tag{3.39}$$

where again each commutator acts on either the first or the second set of spacetime coordinates.

It follows that the condition for the translationally invariant interaction term can be given as

$$[[\partial,\ ],\hat{\mathbf{U}}] = 0. \tag{3.40}$$

In the above equation we treat $[\partial,\ ]$ as a rank-4 operator in its own right, so that commuting it with another rank-4 operator leaves us with a rank-4

50

operator. When we apply this meta-commutator to the Fourier transformed interaction term from the equation (3.37) we are left with

$$(k - l + m - n)U(k, l, m, n) = 0. \tag{3.41}$$

As a constraint on function $U(k, l, m, n)$ the above equation has no non-trivial solutions. However, as a constraint on a kernel of a differential operator the above equation gives us

$$U(k, l, m, n) = V(k, l, m, n)\delta(k - l + m - n). \tag{3.42}$$

We have arrived at the expression for the most general form of translationally invariant interaction term for theories with interactions up to the fourth order in fields. We see that this interaction term is in general a four-point function in momentum space with one overall delta function.

4 Quantization of non-local ϕ^4 theory

4.1 Overview

In the previous chapter we discussed an effective ϕ^4 interaction theory that takes into the account the derivative interactions and we modeled an effective gravitational interaction in this way. We showed that it is possible to model the effective gravitational interaction as a translation-invariant four-point interaction term. In this view, the dimensionful gravitational constant is viewed as a mass parameter in the ϕ^4 interaction potential. However, it does not correspond to the mass of any particle. This would be undesirable since we know that gravitational interaction is long ranged, while massive interaction fields would lead to a short-range interaction.

Another important feature of the effective interaction potential was that we managed to take the derivative terms from the sources and incorporate them into the four-point interaction function. This allowed us to take the terms from the numerators and place them in the denominator of the function, and thus potentially improve the perturbative quantization properties of this theory. In this chapter we'll show how it's possible to construct an integrable perturbative model that in low energy limit yields a gravity-like interaction in addition to the standard local ϕ^4 interaction.

4.2 Feynman Rules

In the model that we build for the purpose of our calculations we'll like to retain all the important features of the gravitational interaction that are characteristic of it, and leave out parts that it shares with other long-range interactions. In particular, this means leaving out all the terms that correspond to $1/r$ potential, and retaining the derivative-coupling terms. For the sake of greater generality and additional simplicity, we will not try to reproduce the GR-like gravitational terms from the previous chapter. Instead we'll write an interaction term that models the gravity-like interaction by the virtue of being a general four-point function whose first non-constant term in Fourier expansion is quadratic in the momenta. In addition to that, the interaction term needs to have a four-point perturbation symmetry in momenta:

$$k_1 \leftrightarrow k_2 \leftrightarrow k_3 \leftrightarrow k_4.$$

The scalar field propagator

$$p \longrightarrow q$$

is given by

$$S(p,q) = f(p^2)\delta(p-q) = \frac{1}{p^2 - m^2}\delta(p-q). \tag{4.1}$$

The four-point interaction vertex

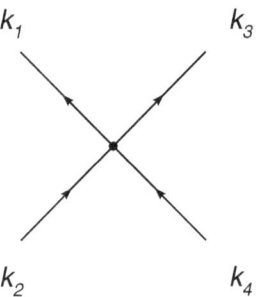

is given by

$$
\begin{aligned}
D(k_1, k_2, k_3, k_4) &= U(k_1, k_2, k_3, k_4)\delta(k_1 - k_2 + k_3 - k_4) \\
&= U(L_1^2(k_1^2 + k_2^2 + k_3^2 + k_4^2) \\
&\quad + L_2^2(k_1 \cdot k_3 + k_1 \cdot k_2 + k_1 \cdot k_4 + k_2 \cdot k_3 + k_2 \cdot k_4 + k_3 \cdot k_4)) \times \\
&\quad \times \delta(k_1 - k_2 + k_3 - k_4) \\
&= \frac{1}{[1 + L_1^2(k_1^2 + k_2^2 + k_3^2 + k_4^2) + L_2^2(k_1 \cdot k_3 + k_1 \cdot k_2 + k_1 \cdot k_4 + k_2 \cdot k_3 + k_2 \cdot k_4 + k_3 \cdot k_4)]^\alpha} \times \\
&\quad \delta(k_1 - k_2 + k_3 - k_4)
\end{aligned}
\tag{4.2}
$$

In the above expression for the four-point interaction term we've chosen the potential function $U(k_1, k_2, k_3, k_4)$ in such a way that in the low-energy limit it can be approximated by

$$U(k_1, k_2, k_3, k_4) \approx 1 - \alpha[L_1^2(k_1^2 + k_2^2 + k_3^2 + k_4^2)$$
$$+ L_2^2(k_1 \cdot k_3 + k_1 \cdot k_2 + k_1 \cdot k_4 + k_2 \cdot k_3 + k_2 \cdot k_4 + k_3 \cdot k_4)] + \ldots \tag{4.3}$$

that is, the first non-constant term is quadratic in momenta. In the low energy limit this term will correspond to the most general derivative-derivative interaction that is symmetrical in terms of the incoming momenta. Since General Relativity can be approximated as a derivative-derivative interaction as well (with an added complication of the inverse distance potential which we choose to ignore in this model), we hope to make a renormalizable theory that can mimic gravity. We see that by taking the derivative terms from the numerator of the interaction to denominator we can avoid the problems that all high-mass-dimension terms encounter in the usual formulation of QFT. [60]

In the most general case, the lenght scales L_1 and L_2 are independant of each other. We recover the usual ϕ^4 theory in the limit where we set them all to 0. In our model they are non-zero but much smaller than other observable langth scales. One can think of them as being at the order of Planck length.

Parameter α serves the purpose of making the denumerator of this potential to be of the right mas dimension. By varying this parameter we can construct a theory that is manifestly renormalizable.

4.3   Self Interaction

The perturbative expansion of the expression for the propagator in ϕ^4 scalar field theory is given by

This expansion can be written in terms of the expressions for the propagators as:

$$S_R(p, q) = S(p, q) + S(p, k_1)D(k_1, k_2, k_3, k_4)S(k_2, k_3)S(k_4, q) + \ldots \tag{4.4}$$

The one loop self interaction term

represents the following calculation:

$$
\begin{aligned}
S^*(p, q) &= D(p, l_1, l_2, q) S(l_1, l_2) \\
&= U(p, q, l_1, l_2) f(l_1^2) \delta(l_1 - l_2) \delta(p - q + l_1 - l_2) \\
&= \int U(p, q, k, k) f(k^2) \delta(p - q) d^d k \\
&= \int \frac{1}{[1 + L_1^2(p^2 + q^2 + 2k^2) + L_2^2(2p \cdot k + 2q \cdot k + p \cdot q + k^2)]^\alpha} \frac{1}{(k^2 - m^2)} d^d k \delta(p - q) \\
&= \int \frac{1}{[1 + L_1^2(2p^2 + 2k^2) + L_2^2(4p \cdot k + p^2 + k^2)]^\alpha} \frac{1}{(k^2 - m^2)} d^d k \delta(p - q) \\
&= \int I(p, k) d^d k \delta(p - q), \\
&= f^*(p^2) \delta(p - q),
\end{aligned}
$$

$$(4.5)$$

where $f^*(p^2) = \int I(p, k) d^d k$.

This gives us the following value of the effective propagator (up to one loop interaction):

$$
S_E(p, q) = \frac{1}{p^2 - m^2 - f^*(p^2)} \delta p - q \tag{4.6}
$$

Now,

$$
I = \frac{1}{A^\alpha B} = \int_0^1 \frac{x^{\alpha-1}}{[xA + (1-x)B]^{\alpha+1}} \frac{\Gamma(\alpha+1)}{\Gamma(\alpha)} dx. \tag{4.7}
$$

If we take $A = L_1^2[M^2 + (2p^2 + 2k^2) + \beta(4p \cdot k + p^2 + k^2)]$ and $B = (k^2 - m^2)$, where $\beta = (L_1/L_2)^2$, $M^2 = 1/L_1^2$, we end up with

$$I(p, k) = \frac{\Gamma(\alpha + 1)}{\Gamma(\alpha)} \frac{1}{L_1^{2\alpha}}$$

$$\times \int_0^1 \frac{x^{\alpha-1} dx}{\{x[M^2 + (2p^2 + 2k^2) + \beta(4p \cdot k + p^2 + k^2)] + (1 - x)(k^2 - m^2)\}^{\alpha+1}}$$

$$= \frac{\Gamma(\alpha + 1)}{\Gamma(\alpha)} \frac{1}{L_1^{2\alpha}}$$

$$\times \int_0^1 \frac{x^{\alpha-1} dx}{\{[(1 - x) + x(2 + \beta)]k^2 + 4\beta x k \cdot p - [(1 - x)m^2 - xM^2 - x(2 + \beta)p^2]\}^{\alpha+1}}$$

$$= \frac{\Gamma(\alpha + 1)}{\Gamma(\alpha)} \frac{1}{L_1^{2\alpha}} \int_0^1 \frac{x^{\alpha-1} dx}{C(x)^\alpha [k^2 + 2k \cdot q - \Delta]^{\alpha+1}},$$

$$(4.8)$$

where

$$q = \frac{2\beta x p}{C(x)},$$

$$\Delta = \frac{(1 - x)m^2 - x[M^2 + (2 + \beta)p^2]}{C(x)},$$

$$(4.9)$$

and

$$C(x) = 1 + x(1 + \beta). \qquad (4.10)$$

Now we have

$$\int \frac{d^d k}{(k^2 + 2k \cdot q - \Delta)^{\alpha+1}} = \frac{i\pi^{d/2}}{\Gamma(\alpha + 1)(-q^2 - \Delta)^{\alpha+1-d/2}} \Gamma(\alpha + 1 - d/2), \quad (4.11)$$

so that

$$f^*(p^2) = \frac{\Gamma(\alpha + 1 - d/2)}{\Gamma(\alpha)} \frac{1}{L_1^{2\alpha}} \times$$

$$\times \int_0^1 \frac{x^{\alpha-1} dx}{C(x)^{d/2}\{-4\beta^2 x^2 p^2 - [1 + x(1 + \beta)][(1 - x)m^2 - x(M^2 + (2 + \beta)p^2)]\}^{\alpha+1-d/2}}$$

$$= \frac{\Gamma(\alpha + 1 - d/2)}{\Gamma(\alpha)} \frac{1}{L_1^{2\alpha}} \int_0^1 \frac{x^{\alpha-1} dx}{[1 + (1 + \beta)x]^{d/2}[m^2 + ax + bx^2]^{\alpha+1-d/2}},$$

$$(4.12)$$

where

$$a = \beta m^2 - M^2 - (2 + \beta)p^2,$$
$$b = (3\beta^2 - 3\beta - 2)p^2 - (1 + \beta)(m^2 + M^2).$$

(4.13)

Gamma function has singularities at 0 and all negative integers. For the values of d and α that make the above gamma functions singular, the integral total integral diverges. nevertheless, we can confine all the divergence to the gamma function and evaluate the finite part of the integrals.

Case 1: $d = 4$, $\alpha = 1$.

This yields a particularly simple form of the integrals:

$$f^*(p^2) = \frac{\Gamma(0)}{\Gamma(1)} \frac{1}{L_1^2} \int_0^1 \frac{dx}{[1 + x(1 + \beta)]^2} = \frac{\Gamma(0)}{\Gamma(1)} \frac{1}{L_1^2} [\log(\beta + 2) + \frac{1}{\beta + 2} - 1].$$

(4.14)

We see that for $\beta = -1$ the finite part of the integral is identically zero, and we have no correction to the propagator at the one-loop level.

Case 2: $d = 4$, $\alpha = 2$.

This is a more complicated and interesting case. For this choice of α and d all integrals will be convergent. We have

$$f^*(p^2) = \frac{\Gamma(1)}{\Gamma(2)} \frac{1}{L_1^4} \int_0^1 \frac{x\,dx}{[1 + x(1 + \beta)]^2(m^2 + ax + bx^2)}$$
$$= \frac{\Gamma(1)}{\Gamma(2)} \frac{1}{L_1^4} J,$$

(4.15)

where

$$J \equiv \int_0^1 \frac{x\,dx}{[1 + x(1 + \beta)]^2(m^2 + ax + bx^2)}$$
$$= \frac{1}{2[m^2(\beta + 1)^2 - a(\beta + 1) + b]^2\sqrt{4bm^2 - a^2}} \times$$
$$\times \left\{ -2[a(m^2(\beta + 1)^2 + b) - 4(\beta + 1)bm^2] \arctan\left(\frac{\sqrt{4bm^2 - a^2}}{2m^2 + a}\right) + \right.$$
$$\left. + \sqrt{4bm^2 - a^2}[m^2(\beta + 1)^2 - b] \log\left(\frac{m^2(\beta + 2)^2}{m^2 + a + b}\right) \right\}$$

(4.16)

for $4bm^2 - a^2 > 0$, and

$$J = \frac{1}{2[m^2(\beta+1)^2 - a(\beta+1) + b]^2\sqrt{a^2 - 4bm^2}} \times$$
$$\times \left\{ \left[a\left(m^2(\beta+1)^2 + b\right) - 4(\beta+1)bm^2\right] \operatorname{arctanh}\left(\frac{\sqrt{a^2 - 4bm^2}}{2m^2 + a}\right) + \right.$$
$$\left. + \sqrt{a^2 - 4bm^2} \left[m^2(\beta+1)^2 - b\right] \log\left(\frac{m^2(\beta+2)^2}{m^2 + a + b}\right) \right\}$$

(4.17)

for $4bm^2 - a^2 < 0$.

We see that in this case we also obtain a finite result for the momentum space integration. However, for this choice of parameters, the integral depens on momentum p as well. (This is because a and b are p-dependant. In order to make sense of the result and make the connection with the original non-perturbed theory, we need to engage in some form of renormalization. Unlike the standard QFT, this will be a finite quantity renormalization.

4.4   Interaction Term

Besides the perturbative expansion of the particle propagator, we need to take into the consideration the perturbative expansion of the interaction term as well. With the notation that we introduced, this is also a rather strightforward, albeit a tedious procedure. In the following section we'll demonstrate the formal mathematical structure of this expansion. Due to the rather unwieldy form of the interaction potential, the actual algebraic form of the momentum space integrals will not be written down.

4.4.1   Interaction Term 1

We are calculating the following term:

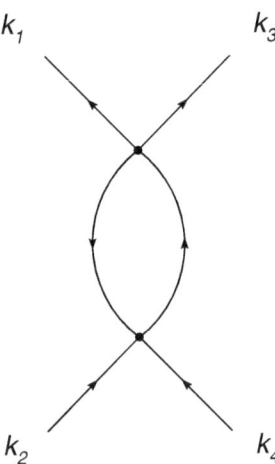

$$D_1^*(k_1, k_2, l_1, l_2) = D(k_1, p, l_1, q)S(p, r)S(q, s)D(r, k_2, s, l_2)$$
$$= U(k_1, p, l_1, q)f(p^2)f(q^2)U(r, k_2, s, l_2)\times$$
$$\times \delta(k_1 - p + l_1 - q)\delta(p - r)\delta(q - s)\delta(r - k_2 + s - l_2)$$
$$= U(k_1, r, l_1, s)f(r^2)f(s^2)U(r, k_2, s, l_2)\times$$
$$\times \delta(k_1 - r + l_1 - s)\delta(r - k_2 + -l_2)$$
$$= U(k_1, r, l_1, l_2 + k_2 - r)f(r^2)f((l_2 + k_2 - r)^2)U(r, k_2, l_2 + k_2 - r, l_2)\times$$
$$\times \delta(k_1 - k_2 + l_1 - l_2)$$
$$= U_1^*(k_1, k_2, l_1, l_2)\delta(k_1 - k_2 + l_1 - l_2),$$

$$(4.18)$$

where

$$U_1^*(k_1, k_2, l_1, l_2) = \int dr U(k_1, r, l_1, l_2 + k_2 - r)f(r^2)f((l_2 + k_2 - r)^2)U(r, k_2, l_2 + k_2 - r, l_2)$$

$$(4.19)$$

4.4.2 Interaction Term 2

We are calculating the following term:

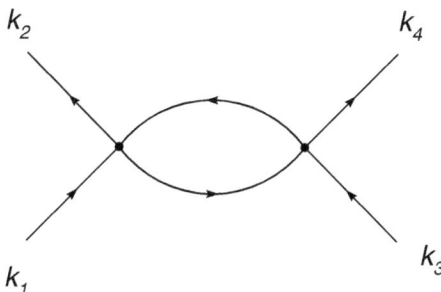

$$D_2^*(k_1, k_2, l_1, l_2) = D(k_1, k_2, p, q)S(p, r)S(q, s)D(r, s, l_1, l_2)$$

$$= U(k_1, k_2, p, q)f(p^2)f(q^2)U(r, s, l_1, l_2) \times$$

$$\times \, \delta(k_1 - k_2 + p - q)\delta(p - r)\delta(q - s)\delta(-r + s + l_1 - l_2)$$

$$= U(k_1, k_2, r, s)f(r^2)f(s^2)U(r, s, l_1, l_2) \times$$

$$\times \, \delta(k_1 - k_2 + r - s)\delta(-r + s + l_1 - l_2)$$

$$= U(k_1, k_2, r, r - l_1 + l_2)f(r^2)f((r - l_1 + l_2)^2)U(r, r - l_1 + l_2, l_1, l_2) \times$$

$$\times \, \delta(k_1 - k_2 + l_1 - l_2)$$

$$= U_2^*(k_1, k_2, l_1, l_2)\delta(k_1 - k_2 + l_1 - l_2),$$

$$(4.20)$$

where

$$U_2^*(k_1, k_2, l_1, l_2) = \int dr\, U(k_1, k_2, r, r - l_1 + l_2)f(r^2)f((r - l_1 + l_2)^2)U(r, r - l_1 + l_2, l_1, l_2)$$

$$(4.21)$$

4.4.3 Interaction term 3

We are calculating the following term:

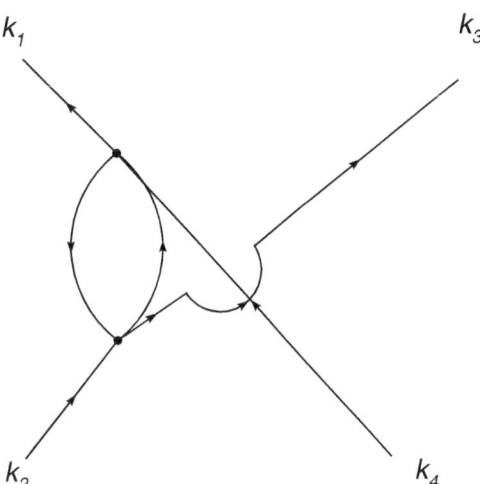

$$D_3^*(k_1, k_2, l_1, l_2) = D(k_1, p, -l_2, -q)S(p, r)S(q, s)D(r, k_2, -s, -l_1)$$
$$= D(k_1, p, -l_2, -q)f(p^2)f(q^2)U(r, k_2, -s, -l_1) \times$$
$$\times \delta(k_1 - p - l_2 + q)\delta(p - r)\delta(q - s)\delta(r - k_2 - s + l_1)$$
$$= U(k_1, r, -l_2, -s)f(r^2)f(s^2)U(r, k_2, -s, -l_1) \times$$
$$\times \delta(k_1 - r - l_2 + s)\delta(r - k_2 - s + l_1)$$
$$= U(k_1, r, -l_2, -r - k_2 + l_1)f(r^2)f((r + k_2 - l_1)^2)U(r, k_2, r + k_2 - l_1, -l_1) \times$$
$$\times \delta(k_1 - k_2 + l_1 - l_2)$$
$$= U_3^*(k_1, k_2, l_1, l_2)\delta(k_1 - k_2 + l_1 - l_2), \tag{4.22}$$

where

$$U_3^*(k_1, k_2, l_1, l_2) = \int dr U(k_1, r, -l_2, -r - k_2 + l_1)f(r^2)f((r + k_2 - l_1)^2)U(r, k_2, r + k_2 - l_1, -l_1) \tag{4.23}$$

We have the following form of the effective interaction (up to one loop):

This Feynman diagram has the following form in terms of expressions for field propagators and vertex functions:

61

$$D_E(k_1, k_2, l_1, l_2) = D(k_1, k_2, l_1, l_2)$$
$$+ D_1^*(k_1, k_2, l_1, l_2) + D_2^*(k_1, k_2, l_1, l_2) + D_3^*(k_1, k_2, l_1, l_2)$$

$$(4.24)$$

5 Non-Local Gauge Invariance in Physical Theories II: Quantum Theory

5.1 Attempts to Quantize GR

Historically there have been two main approaches to the problem of quantization of gravity: the covariant approach and the canonical approach [36, 37, 38, 39, 40]. In the covariant approach we treat the gravitational field as just another relativistically invariant field and then we apply the perturbative quantization methods that have been developed for all other covariant field theories (QED, QCD, etc.). However, there are two main reason why this method should not work for the GR. The coupling constant for the GR is Newton's constant, and this constant is not dimensionless. This means that at each order of perturbative expansion we would need additional higher order terms in the Lagrangian in order to absorb the renormalized quantities that come out of the perturbation theory. Furthermore, the GR Lagrangian contains the non-polynomial interactions of the gravitational field, so for each additional level of the perturbation we would need new interaction terms. All of this renders GR a non-renormalizable theory in the strictest sense. However, a hope existed for a while that due to some fortuitous cancellations the non-renormalizable terms may cancel each other out at each level of perturbation. Indeed this happens in GR at one loop level, but further studies have shown that there are no such cancellations at higher orders. With the introduction of supersymmetry the cancellations can be achieved at two-loop level as well, but that is as far as it goes.

The canonical approach is the program of defining a Schrödinger-like equation for the gravitational field. In this approach we foliate the space-time into three-dimensional hypersurfaces and then we define a Hamiltonian that tells us how the three-dimensional geometry evolves from one hypersurface to another. When we deal with the classical field equations this approach breaks the explicit relativistic invariance, but the equations themselves are valid field equations. When we try to quantize these equations an interesting thing happens: the resulting equation (Wheeler-DeWitt equation) has no time dependence. Conceptually we can understand this as follows: in GR time translation is just another form of gauge transformation, and it is not present in the gauge-invariant quantum equations. This would be the coordinate time, which is distinct from the physical time. The actual physical time needs to be extracted from the solution to the equations. This is a daunting task even for the simplest model geometries. This problem is known as the problem of time in the Canonical Quantum Gravity. In this paper we deal only with the covariant approach to the quantization and we will not address this problem.

One area where the Quantum Field Theory and GR have met with a reasonable success is the Quantum Field Theory in Curved Space-time [31, 33]. There one deals with Quantum Fields that propagate in a fixed classical space-time. Some of the most interesting results in this field (Hawking Radiation, Unruh Effect) have motivated the research in Quantum Gravity over the past three decades. However, since the gravitational equations are treated classically, the theory doesn't tell us anything about the back-reaction of the Quantum Fields on Gravity.

Today the two main programs that work on Quantum Gravity are the String Theory and Loop Quantum Gravity. These are two complex theoretical frameworks and it is beyond the scope of this paper to review them. It is important to emphasize that both of those programs view GR as a low energy limit of some more fundamental theory. In our approach there is no underlying theory from which we derive the low-energy field equations, but our theory does have an interesting new high-energy structure.

5.2 Quantization of the Non Local Gauge Field Theory

In this section we will show how to derive Feynman rules for the perturbative quantization of the non-local gauge field theory. Since we have defined the theory in the position space, all the Feynman rules will also be in position space. We would not gain any algebraic simplification if we were to work in momentum space, because the most general non-local operators are not diagonalizable in momentum space either.

64

Since our entire discussion of the classical theory was based on the definition of the theory in terms of Action, the most straightforward quantization method is in terms of Functional Integrals. The Functional Integral is defined as

$$Z = \int \mathcal{D}\psi \mathcal{D}\hat{A}^\mu e^{-i<\psi, \not{D}\psi> - \mathrm{itr}\hat{F}^{\mu\nu}\hat{F}_{\mu\nu}}. \tag{5.1}$$

This Functional Integral gives us the Partition function in terms of which we can calculate the correlation functions of various fields. However, some caution needs to be used when dealing with traces of differential operators. In general, these traces are infinite, and at intermediate steps we need to use some normalization prescription. However, since we are only interested in quantities of the form

$$< \hat{A}^\mu \hat{A}^\nu > = \frac{\int \mathcal{D}\psi \mathcal{D}\hat{A}^\alpha \hat{A}^\mu \hat{A}^\nu e^{-i<\psi, \not{D}\psi> - \mathrm{itr}\hat{F}^{\delta\gamma}\hat{F}_{\delta\gamma}}}{\int \mathcal{D}\psi \mathcal{D}\hat{A}^\alpha e^{-i<\psi, \not{D}\psi> - \mathrm{itr}\hat{F}^{\delta\gamma}\hat{F}_{\delta\gamma}}}, \tag{5.2}$$

any such prescription will yield finite correlation functions, as long as it is self-consistent and gauge invariant.

The requirement of gauge invariance of the partition function means that we need to modify expression (59). We need to introduce a Haar measure over the unitary operators. This is accomplished through gauge-fixing and introduction of ghost fields.

We choose the non-local generalization of the Lorentz gauge condition

$$[\partial^\mu, \hat{A}_\mu] = \hat{\omega}, \tag{5.3}$$

where we give $\hat{\omega}$ a Gaussian weight. The result of this gauge fixing is that we need to introduce a functional determinant in the path integral (59). Following Fedeev and Popov we choose to represent this determinant as a functional integral over a new set of anticommuting fields:

$$\det\left([\partial^\mu, \hat{D}_\mu]\right) = \int \mathcal{D}c \mathcal{D}\bar{c} e^{-i\mathrm{tr}[\bar{c}(\partial^\mu, \hat{D}_\mu)c]}. \tag{5.4}$$

The final Action, including all of the effects of Fedeev-Popov gauge fixing, is

$$S = <\psi, (i \not{D} - m)\psi> - i\text{tr}\hat{F}^{\mu\nu}\hat{F}_{\mu\nu} + \frac{1}{2}\xi\text{tr}([\partial^\mu, \hat{A}_\mu]) + <\bar{c}, [\partial^\mu, D_\mu]c> .$$

$$(5.5)$$

5.2.1 propagators

Fermion Propagator

$$p \longrightarrow q$$

In terms of this Action the free-field propagator for the spinorial fields is

$$<\psi(x)\bar{\psi}(y)> = \int \frac{d^4q}{(2\pi)^4} \frac{i(\not{q} + m)}{q^2 - m^2 + i\epsilon} e^{-iq(x-y)},$$

$$(5.6)$$

which in the momentum space reads as

$$S(q, p) = \frac{i(\not{q} + m)}{q^2 - m^2 + i\epsilon} \delta(q - p).$$

$$(5.7)$$

We will mostly use the momentum space form of the propagator since it has a much simpler and more transparant structure. It is also much easier to do calculations with it in this form.

Gauge Field Propagator

$$\mu, \ k_1, k_2 \quad \wedge\!\wedge\!\wedge\!\wedge\!\wedge\!\wedge \quad \nu, \ k_3, k_4$$

For the gauge fields we have

$$< \hat{A}^\mu \hat{A}^\nu > = < \sum_{k=0}^{\infty} A^\mu_{(k)}(x) \mathcal{D}^{(k)} \sum_{l=0}^{\infty} A^\nu_{(l)}(y) \mathcal{D}^{(l)} >$$

$$= \sum_{k=0}^{\infty} < A^\mu_{(k)}(x) A^\nu_{(k)}(y) > \mathcal{D}^{(k)} \mathcal{D}^{(k)}$$

$$= < A^\mu(x) A^\nu(y) > + < A^\mu_\alpha(x) A^\nu_\beta(y) > \partial_1^\alpha \partial_2^\beta + \dots$$

$$= \int \frac{d^4 q}{(2\pi)^4} \frac{-i\eta_{\mu\nu}}{q^2 + i\epsilon} e^{-iq(x-y)} \left[1 - \partial_1^\alpha \partial_{\alpha 2} + (\partial_1^\alpha \partial_{\alpha 2})^2 + \dots \right] \quad (5.8)$$

$$= \int \frac{d^4 q}{(2\pi)^4} \frac{-i\eta_{\mu\nu}}{q^2 + i\epsilon} e^{-iq(x-y)} \sum_{k=0}^{\infty} (-\partial_1^\alpha \partial_{\alpha 2})^k$$

$$= \int \frac{d^4 q}{(2\pi)^4} \frac{-i\eta_{\mu\nu}}{q^2 + i\epsilon} e^{-iq(x-y)} \frac{1}{1 + \partial_1^\alpha \partial_{\alpha 2}}$$

In this derivation we have not used the the gauge-fixing procedure, and we have used the derivative expression for the propagator. This seems rather awkward, but it's good to remember that partial derivatives correspond to momenta in the momentum space representation, adn there is nothing unusual about having momenta in the denumerator of any given expression. In order to see this more clearly, we will write down the gauge field propagator in the momentum space. We have:

$$< \hat{A}^\mu(k_1, k_2) \hat{A}^\nu(k_3, k_4) > =$$

$$D^{\mu\nu}(k_1, k_2, k_3, k_4) = \frac{-i}{(k_1 - k_2)^2 + i\epsilon}$$

$$\times \left(\eta_{\mu\nu} - (1 - \zeta) \frac{(k_1 - k_2)^\mu (k_1 - k_2)^\nu}{(k_1 - k_2)^2} \right) \quad (5.9)$$

$$\times \frac{1}{1 + \frac{k_1 \cdot k_3}{\alpha M_{Pl}^2}} \delta(k_1 - k_2 + k_3 - k_4),$$

where α is the fine structure constant and M_{Pl} is the Planck's mass. This propagator has not been gauge fixed, so we have retained a gauge-fixing term ζ. In actual calculations we usually choose a specific value for this term. For instance, $\zeta = 0$ is known as Landau gauge, and $\zeta = 1$ is Feynman gauge.

An important thing to notice is that this propagator has two more powers of momentum in the denominator than the Maxwell field propagator. This implies that all of the higher order Feynman diagrams yield finite values, and therefor the theory is superficially renormalizable. We will show how this works in practice in a few example that we'll provide later on.

The explicit introduction of the Planck mass in this propagator serves two purposes: it explicitly shows that this propagator has the same mass dimension as the corresponding propagator in QED, and it helps us see how we can recover the QED propagator by letting $M_{Pl} \to \infty$.

Ghost Propagator

The ghost field propagator is unchanged from the local theory:

$$p \;\text{-----------}\; q \qquad C(p,q) = \frac{1}{p^2 + i\epsilon}\delta(p-q)$$

5.3 Interaction terms

Since the non-local unitary group is not a Lie group, we can not define higher order interaction terms in terms of structure constants of the Lie algebra. Instead, we need to write all the commutators explicitly as higher order tensors. For instance, the commutator of two matrices (rank two tensors) will be a rank six tensor in terms of matrix indecies. It is a multi-linear function that takes two matrices and gives out a third. Let A_{ab} and B_{cd} be two matrices. The commutator matrix is given by

$$K_{ij} = A_{ik}B_{kj} - B_{ik}A_{kj}, \tag{5.10}$$

from where we get for the commutator

$$C_{ijabcd} = \delta_{ai}\delta_{bc}\delta_{dj} - \delta_{ci}\delta_{da}\delta_{bj}. \tag{5.11}$$

For the sake of simplicity we will continue to work with the matrix representation of operators. The final values will be easily obtained by substituting Dirac delta functions for Kronnecker delta functions.

We construct the following tensor:

68

$$F_{ij}^{\mu\nu} = C_{ijabcd}D^{\mu ab}D^{\nu cd}$$
$$= (\delta_{ai}\delta_{bc}\delta_{dj} - \delta_{ci}\delta_{da}\delta_{bj})D^{\mu ab}D^{\nu cd}$$
$$= D_{ib}^{\mu}D_{cj}^{\nu}\delta_{bc} - D_{id}^{\nu}D_{aj}^{\mu}\delta_{da}$$
$$= D_{ib}^{\mu}D_{bj}^{\nu} - D_{id}^{\nu}D_{dj}^{\mu}, \tag{5.12}$$

or

$$\hat{F}^{\mu\nu} = [\hat{D}^{\mu}, \hat{D}^{\nu}], \tag{5.13}$$

which is the standard definition of the field strength for gauge fields. The action for this field strength is given by

$$S = \mathrm{tr}[\hat{F}^{\mu\nu}, \hat{F}_{\mu\nu}]$$
$$= F_{ij}^{\mu\nu}F_{\mu\nu ij}$$
$$= C_{ijabcd}C_{jipqrs}D^{\mu ab}D^{\nu cd}D_{\mu pq}D_{\nu cd}D_{\nu rs}$$
$$= K_{abcdpqrs}\eta_{\mu\nu}\eta_{\lambda\rho}D^{\mu ab}D^{\lambda pq}D^{\nu cd}D^{\rho rs}, \tag{5.14}$$

where

$$K_{abcdpqrs} = C_{ijabcd}C_{jipqrs}$$
$$= (\delta_{ai}\delta_{bc}\delta_{dj} - \delta_{ci}\delta_{da}\delta_{bj})(\delta_{pj}\delta_{qr}\delta_{si} - \delta_{rj}\delta_{sp}\delta_{qi}) \tag{5.15}$$
$$= (\delta_{as}\delta_{dp}\delta_{bc}\delta_{qr} - \delta_{sc}\delta_{bp}\delta_{da}\delta_{qr} - \delta_{rd}\delta_{aq}\delta_{bc}\delta_{sp} + \delta_{qc}\delta_{br}\delta_{da}\delta_{sp}).$$

5.3.1 Four boson vertex

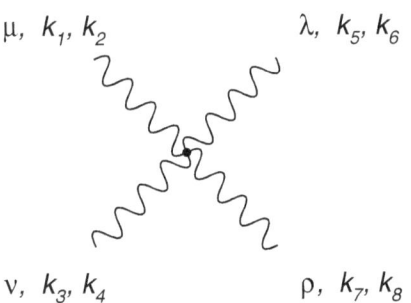

μ, k_1, k_2 λ, k_5, k_6

ν, k_3, k_4 ρ, k_7, k_8

69

The four-boson vertex for the corresponding quantum theory is given by

$$
\begin{aligned}
i\Gamma^{abcdpqrs}_{\mu\nu\lambda\rho} = ig^2[& K^{abcdpqrs}(\eta_{\mu\lambda}\eta_{\nu\rho} - \eta_{\mu\rho}\eta_{\nu\lambda}) \\
& + K^{abpqcdrs}(\eta_{\mu\nu}\eta_{\lambda\rho} - \eta_{\mu\rho}\eta_{\lambda\nu}) \\
& + K^{abrspqcd}(\eta_{\mu\lambda}\eta_{\rho\nu} - \eta_{\mu\nu}\eta_{\rho\lambda})]
\end{aligned}
\tag{5.16}
$$

5.3.2 Three boson vertex

μ, k_1, k_2

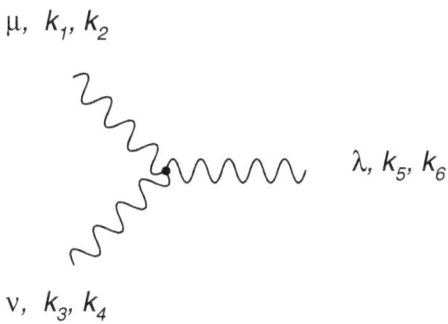

λ, k_5, k_6

ν, k_3, k_4

We continue with the 'discrete' form of the non-local theory that we used in the previous section to derive the form of the four-boson vertex. In this theory the three boson interaction will have the following form

$$
S_3 = K_{abcdpqrs}\partial^{\mu ab}A^{\nu cd}A_{\mu pq}A_{\nu cd},
\tag{5.17}
$$

where now $\partial^{\mu ab}$ and $A^{\nu cd}$ are the discrete versions of the partial derivative operator and gauge field operator respectively. This interaction term leads to the following momentum-space form of the discret three-boson quntum vertex:

$$
\begin{aligned}
i\Gamma^{abcdpq}_{\mu\nu\lambda} = ig(& K^{abcdijpq}k^{\lambda}_{ij}\eta^{\mu\nu} \\
& + K^{cdpqijab}k^{\mu}_{ij}\eta^{\nu\lambda} \\
& + K^{pqabijcd}k^{\nu}_{ij}\eta^{\mu\lambda}),
\end{aligned}
\tag{5.18}
$$

where g is the interaction constant, and k^{μ}_{ij} is the discrete form of the momentum operator.

5.3.3 Other diagrams

The form of all other Feynman diagrams (like vertex factors) is essentially the same as in the case of local gauge field theories. All these diagrams are written explicitly in the Appendix D.

5.4 Perturbative calculations

In this section we will apply the quantization rules to some perturbative QFT calculations. One of the aesthetic advantages of writing all the propagators and vertices in the non-local formalism is that we can collapse a few different computaton rules into a single rule of "contracting the momenta", that is, integrating over the repeated momenta in the given expression. This is analogous to the contraction of indexies in the case of working with tensors. This will become clearer as we work out specific examples.

5.4.1 Tree diagrams

In order to better understand the form of interaction of the non-local gauge theory, we'll take a look at the scattering amplitude for a simple tree-level process. The leading-order contribution is

$$
\begin{aligned}
i\mathcal{M} &= (-ie)^2 \bar{u}(k_2)\gamma^\mu u(k_1) D_{\mu\nu}(k_1, k_2, k_3, k_4)\bar{u}(k_4)\gamma^\nu u(k_3) \\
&= (-ie)^2 \bar{u}(k_2)\gamma^\mu u(k_1)\frac{-i\eta_{\mu\nu}}{(k_1 - k_2)^2 + i\epsilon} \\
&\quad \times \frac{1}{1 + \frac{k_1 \cdot k_3}{e^2 M_{Pl}^2}}\delta(k_1 - k_2 + k_3 - k_4)\bar{u}(k_4)\gamma^\nu u(k_3)
\end{aligned}
\tag{5.19}
$$

where $u(p)$ is an external fermion line, and we've used a gauge-fixed form of the inteaction (Feynman gauge). In the non-relativistic limit the above expression reduces to

$$
i\mathcal{M} \approx \frac{-ie^2}{|\mathbf{k_2} - \mathbf{k_1}|^2}\frac{1}{1 + \frac{m_1 m_2}{e^2 M_{Pl}^2}}
\tag{5.20}
$$

71

which implies the following form for the low-energy potential function:

$$V(r) = \frac{\alpha}{r} \frac{1}{1 + (G_N/\alpha)m_1 m_2} \approx \frac{\alpha}{r} \left(1 - (G_N/\alpha)m_1 m_2 + [(G_N/\alpha)m_1 m_2]^2 - \ldots\right)$$

(5.21)

where the approximate expression is valid fro small values of $(G_N/\alpha)m_1 m_2$. We see that in this expansion the gravitational interaction comes as a second order correction to the Coulomb potential. Furthermore, the equation also implies that there would be higher order corrections as well. However, since we don't observe these interaction effects, we would like to find out if the theory implies that they are indeed negligable at the ordinary energy scales. We can see that this is indeed the case by observing that, for instance, $(G_N/\alpha)m_1 m_2$ factor is really small, and the strength of the first post-gravitational interaction term as compared to gravity is as weak as the gravity is compared to electromagnetic interaction.

5.4.2 The Fermion Self-Interaction

One of the main attractions of the non-local gauge theory is that the form of the Feynman diagrams naturally follows from the algebraic structure of the propagators. This follows from the fact that we treat fields as vectors in a functional vector space, so that propagators become a multi-linear functions (tensors) over that same vector space. In particular, in terms of its functional description fermion propagator is a rank-2 tensor, and gauge field propagator is a rank-4 tensor. When viewed that way, all Feynman diagram prescrptions for integrating over momenta can be reduced to a rule of contracting "indices". For instance, the electron two-point function can be expanded in the following way:

$$S^*(p, q) = S(p, q) + S(p, k_1)\gamma^\mu D_{\mu\nu}(k_1, k_2, k_3, k_4)\gamma^\nu S(k_2, k_3)S(k_4, q) + \ldots,$$

(5.22)

where we integrate over all repeated momenta. In the case of the usual form of the QED propagator, this will result in the usual QED integral. In terms of Feynman diagrams, the above equation correspnds to the following diagram:

The non-trivial part of the above equation is

$$-i\Sigma_2(k_1, k_4) = (-ie)^2\gamma^\mu D_{\mu\nu}(k_1, k_2, k_3, k_4)\gamma^\nu S(k_2, k_3) \qquad (5.23)$$

Which corresponds to the following Feynman diagram:

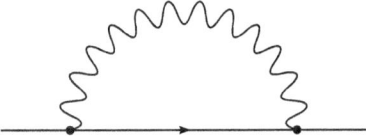

We have

$$\begin{aligned}
\Sigma_2(k_1, k_4) &= -ie^2\gamma^\mu D_{\mu\nu}(k_1, k_2, k_3, k_4)\gamma^\nu S(k_2, k_3) \\
&= -e^2\gamma^\mu \frac{-i\eta_{\mu\nu}}{((k_1 - k_2)^2 + i\epsilon)(1 + \frac{k_1 \cdot k_3}{\alpha M_{Pl}^2})}\delta(k_1 - k_2 + k_3 - k_4)\gamma^\nu \\
&\quad \times \frac{i(\slashed{k}_2 + m)}{k_2^2 - m^2 + i\epsilon}\delta(k_2 - k_3) \\
&= -e^2\int\frac{d^4k}{(2\pi)^4}\gamma^\mu\frac{-i\eta_{\mu\nu}}{((k_1 - k)^2 + i\epsilon)(1 + \frac{k_1 \cdot k}{\alpha M_{Pl}^2})}\gamma^\nu \\
&\quad \times \frac{i(\slashed{k} + m)}{k^2 - m^2 + i\epsilon}\delta(k_1 - k_4)
\end{aligned}$$

$$(5.24)$$

The divergence of the above integral is determined by the number of momenta that we have in the denumarator. In QED we did not have the extra

$$\frac{1}{1 + \frac{k_1 \cdot k}{\alpha M_{Pl}^2}}$$

term, and the resulting integral was logorithmically divergent. We see that this term tends to 1 as $M_{Pl} \to \infty$, in which case we recover the QED result. However, for any finite value of M_{Pl} the integral above will be convergent. We still need to perform a finite renormalization procedure, that is renormalization of masses and other parameters of the theory just like in case of

QED, but with only diference that these quantities will be renormalized by a finite amount.

5.4.3 Fermion-gauge boson vertex

The perturbative expansion of the fermion - gauge boson vertex has the form:

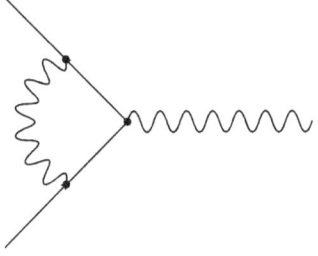

Vertex correction 1

We are interested in the following interaction diagram:

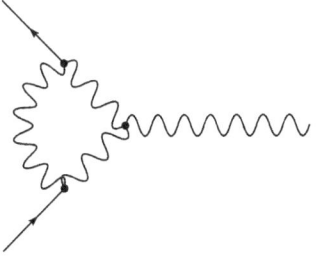

This diagram represents the interaction that was already present in the local gauge field theory. However, the presence of non-local terms will alter the value of this diagram.

Vertex correction 2

We are interested in the following interaction diagram:

74

This diagram is reminiscent of the similar diagram that is present in non-abelian gauge field theories. It arises here because of the non-commutativity of the non-local gauge fields.

5.4.4 Gauge boson self interaction

The perturbative expansion of the gauge boson propagator has the form:

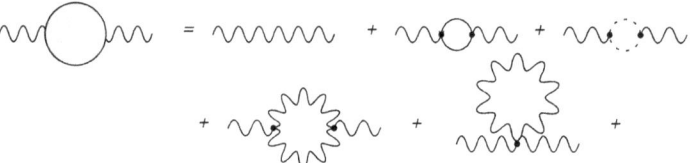

Fermion bubble

We are looking at the following diagram:

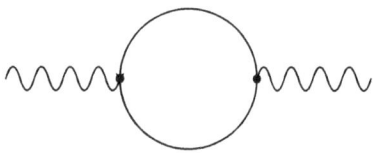

The expression for it is given by

$$D^{\mu\alpha}(k_1, k_2, l_1, l_2)\text{tr}\left[\gamma_\alpha S(l_1, l_3)\gamma_\beta S(l_2, l_4)\right]D^{\beta\nu}(l_3, l_4, k_3, k_4) \qquad (5.25)$$

Ghost bubble

We have the following diagram:

The expression for it is given by

$$D_{\mu\alpha}(k_1, k_2, l_1, l_2)(l_1^\mu - l_2^\mu)\delta(l_1 - l_3)\delta(l_2 - l_4)C(l_3, l_5)C(l_4, l_6)$$
$$\times (l_5^\mu - l_7^\mu)\delta(l_5 - l_7)\delta(l_6 - l_8)D_{\nu\alpha}(l_7, l_8, k_3, k_4) \qquad (5.26)$$

Gauge boson bubble 1

We have the following diagram:

The expression for this diagram is given by

$$D_{\mu\alpha}(k_1, k_2, l_1, l_2)\Gamma^{\alpha\beta\gamma\delta}(l_1, l_2, l_3, l_4, l_5, l_6, l_7, l_8)D_{\beta\gamma}(l_3, l_4, l_5, l_6)D_{\delta\nu}(l_7, l_8, k_3, k_4) \tag{5.27}$$

Gauge boson bubble 2

We have the following diagram:

The expression for this diagram is given by

$$D_{\mu\alpha}(k_1, k_2, l_1, l_3)\Gamma^{\alpha\beta\gamma}(l_1, l_2, l_3, l_4, l_5, l_6)D_{\beta\lambda}(l_3, l_4, l_7, l_8)D_{\gamma\sigma}(l_5, l_6, l_9, l_{10})$$
$$\times \Gamma^{\lambda\sigma\rho}(l_7, l_8, l_9, l_{10}, l_{11}, l_{12})D_{\rho\nu}(l_{11}, l_{12}, k_3, k_4) \tag{5.28}$$

In all of the above expressions we integrate ove the repeated momenta. As mentioned earlier, this is analogous to contraction of indecies for tensors. By writing down all gauge propagators as four-point functions, and all vertex factors as many-point functions, it becomes transparent how to contract different elements in order to form perturbative higher order integrals. The eventual simplification of the multiple integrals comes around because the propagators and vertecies contain delta functions. The actual calculations are rather involved, and will not be attempted at this point. The reason why we believe that they lead to renormalzable terms is because they will also containhigher orders of momenta in the denominator as compared to the local gauge field theory case. We've already seen how this comes about in the case of the fermion self-interaction.

6 Gauging of the Spin Degrees of Freedom

6.1 Introduction

Traditionally it is thought that the fundamental gauge groups serve two purposes in Physics. The Poincare Group is the fundamental group for all physical theories and it determines the kinematic content of the theory. All physical quantities should be covariant under this group. If we think of physical fields as vectors in a Hilbert space, then Poincare Group acts on spacetime and spin indices of these vectors.

On the other hand we have a group of gauge degrees of freedom which in the Standard Model determines dynamics. It is due to these groups that we have fundamental interactions and forces. The forces of the standard model correspond to $U(1) \times SU(2) \times SU(3)$ gauge group. Not all elementary particles are subject to all the fundamental forces. Charged leptons interact through electromagnetic and weak force, neutrinos only through weak force, and quarks through electromagnetic, weak and strong force. The gauge degrees of freedom are independent of the PG degrees of freedom, and the fields on which the gauge group acts have another set of indices. As of writing of this paper there is no universally accepted overarching gauge group that incorporates all of these forces, although the group $SU(5)$ has been considered one of the better candidates.

This view of the independence of the Poincaré Group and the gauge group becomes murky once we try to incorporate gravity into this picture. The universally accepted theory of gravity is General Relativity (GR). In its original formulation GR is thought of as a theory of local coordinate transformations - diffeomorphisms. These transformations act on the coordinates of fields, but there is no *a priori* connection between the group of diffeomorphisms and PG. Another approach is to define the gauge group of GR to be the group of local Lorentz transformations. In this approach The gauge group itself acts on spacetime and spin degrees of freedom.

In the earlier chapters I outlined an approach to formulating gravity as a non-local gauge field theory (NLGFT). This approach had an advantage of being similar in spirit to the Standard Model approach to gauge freedom. We formulated gauge transforms in terms of non-local unitary transformations of the physical fields. This was an eminently algebraic approach, and it corresponded to a multiplication of a vector with a matrix. This contrasts with the view in which the gauging is defined in terms of the parallel transport on the underlying manifold. This approach regarded the spacetime degrees of freedom and the local gauge group degrees of freedom in an unified way. What was left out were the spin degrees of freedom. In this chapter I will try to incorporate those degrees of freedom as well.

6.2   A Simple 2-spinor Model

In this section I will introduce a simple 2-spinor model that has all the basic properties of the spin gauge group. I will show how the requirements of **local** spin gauge invariance and the minimal coupling impose the introduction of the NLGFT.

6.2.1   Local $U(1)$ gauge invariance.

I will first review the local $U(1)$ gauge invariance. Let $\psi(x)_a$ be a massless bispinnor ($a = 1, 2$). Then the action for this field is given by

$$S = \int \psi(x)_a^* \; \not{\partial}^{ab} \psi(x)_b dx, \tag{6.1}$$

where $\not{\partial} = \sigma^\mu \partial_\mu$, $\sigma_0 = I$, $\sigma_a, a = 1, 2, 3$ are the Pauli sigma matrices, and we have used the summation over the repeated indices. We would like to require that action S is invariant under a local $U(1)$ gauge transformation of thefield $\psi(x)$. The field $\psi(x)$ transforms as

$$\psi(x) \rightarrow U(x)\psi(x), \tag{6.2}$$

where $U(x) = \exp[i\theta(x)]$, and *theta*(x) is an arbitrary suitably well behaved function. Under this transformation the derivative term in action S transforms as

$$\partial\!\!\!/\psi(x) \quad \rightarrow \quad \partial\!\!\!/\left[e^{i\theta(x)}\psi(x)\right] \tag{6.3}$$

$$= \quad e^{i\theta(x)}\left[\partial\!\!\!/\psi(x) + i\ \partial\!\!\!/\theta(x)\psi(x)\right] \tag{6.4}$$

$$= \quad e^{i\theta(x)}\ \partial\!\!\!/\psi(x) + ie^{i\theta(x))}\ \partial\!\!\!/\theta(x)\psi(x). \tag{6.5}$$

We see that the total action would be invariant under the local $U(1)$ transformation were it not for the last term in the above equation. In order to get rid of that term we need to introduce a gauge field $A^\mu(x)$ whose action is invariant under the following transformation:

$$A^\mu(x) \rightarrow A^\mu(x) + \partial^\mu\theta(x).$$

We accomplish this by introducing a covariant derivative $\mathcal{D}^\mu = \partial^\mu + iA^\mu(x)$. If we want to make the total action invariant, then we would need to introduce action part for this field strength. There is a simple way of doing this, but we'll come to this point later.

6.2.2 Local $U(2)$ Spin Gauge Invariance

Important thing to notice is that σ^μ are the generators of the unitary group $U(2)$. In view of that it is meaningful to define a unitary operator

$$\widehat{U} = \exp\left(i\sigma^\mu\theta_\mu(x)\right),$$

where $\theta_\mu(x)$ are arbitrary local functions. With this definition we deffine a unitary transformation of spinnor ψ as

$$\psi \rightarrow \widehat{U}\psi.$$

We would like to have the total action remain invariant under this transformation. If this were just an ordinary gauge theory, then we would have to worry only about the noncommutativity of the differential operator with the local functions. As it is, we have to take care of the spin indices as well. It will be sufficient for now to deal only with the infinitesimal form of the operator $\widehat{U}$. We write

$$\widehat{U} \approx I + i\sigma^{\mu}\theta_{\mu}(x) = I + i\,\widehat{\theta}.$$

Under this transformation the derivative term in action changes as

$$\begin{align}
\partial\!\!\!/\psi &\rightarrow \partial\!\!\!/\widehat{U}\psi & (6.6)\\
&\approx \partial\!\!\!/(I + i\,\widehat{\theta}\psi) & (6.7)\\
&= \sigma^{\nu}\partial_{\nu}(I + i\sigma^{\mu}\theta_{\nu})\psi & (6.8)\\
&= (\partial\!\!\!/ + i\,\partial\!\!\!/\,\widehat{\theta})\psi & (6.9)\\
&= (\partial\!\!\!/ + i\,\widehat{\theta}\,\partial\!\!\!/ + i[\partial\!\!\!/, \widehat{\theta}])\psi & (6.10)\\
&= (I + i\,\widehat{\theta})\,\partial\!\!\!/\psi + i[\partial\!\!\!/, \widehat{\theta}]\psi. & (6.11)
\end{align}$$

We observe that the noncommutativity of $\partial\!\!\!/$ and $\widehat{U}$ comes from the fact that on general $[\partial\!\!\!/, \widehat{\theta}] \neq 0$ even when $\widehat{U}$ corresponds to general *global* operator. This is due to the noncommutativity of Pauli sigma matrices. In order to make the entire action invariant under these transformations, we will need to introduce a gauge field that absorbs the extra term. However, unlike the $U(1)$ case, the gauge filed that we introduce could not be entirely a local or global operator, even when $\widehat{\theta}$ is an entirely global operator. This will lead to an introduction of a non-local gauge field. First, let us compute the commutator $[\partial\!\!\!/, \widehat{\theta}]$.

$$[\partial\!\!\!/, \widehat{\theta}] = \sigma^{\mu}\sigma^{\nu}\partial_{\mu}\theta_{\nu} + \sigma^{\mu}\sigma^{\nu}\theta_{\nu}\partial_{\mu} - \sigma^{\nu}\sigma^{\mu}\theta_{\mu}\partial_{\nu}. \tag{6.12}$$

Now, for an arbitrary tensor $D_{\mu\nu}$ we have

$$\begin{align}
\sigma^{\mu}\sigma^{\nu}D_{\mu\nu} &= \sigma^{0}\sigma^{\nu}D_{0\nu} + \sigma^{i}\sigma^{\nu}D_{i\nu} & (6.13)\\
&= \sigma^{0}\sigma^{0}D_{00} + \sigma^{i}D_{0i} + \sigma^{i}\sigma^{j}D_{ij} + \sigma^{i}\sigma^{0}D_{i0} & (6.14)\\
&= D_{00} + D_{ii} + \sigma^{i}(D_{0i} + D_{i0}) + i\sigma_{k}\epsilon^{ijk}D_{jk}, & (6.15)
\end{align}$$

so that

$$[\partial\!\!\!/, \widehat{\theta}] = \sum_{\mu}\partial_{\mu}\theta_{\mu} + \sum_{i}\sigma_{i}(\partial_{0}\theta_{i} + \partial_{i}\theta_{0}) + i\sum_{ijk}\sigma_{k}(\epsilon_{ijk}\partial_{i}\theta_{j}) \tag{6.16}$$

As expected, the derivative operator does not commute with the infinitesimal transformations. In order to remedy this we would need to introduce a covariant derivative, which would consist of the derivative term $\not\partial$ plus a gauge field $\not A$. However, the extra terms that we got after the commutation of the $\not\partial$ with $\hat{\not\partial}$ also contained partial derivatives. Thus, the gauge field that can absorb these terms, also needs to be a derivative-valued field. Which brings us to the following important conclusion: if we try to gauge the spin degrees of freedom, even for the global transformations, we need to introduce a non-local gauge fields.

One thing that we notice with this model is that the gauge transformations for simple 2-spinors are not manifestly relativistically invariant. In order to remedy this, in the next section we will start working with 4-spinors and Dirac's gamma matrices.

6.3   Dirac-unitary transformation

The Dirac-unitary transformation is a linear matrix transformation in four dimensions for which the following condition holds:

$$S^\dagger \gamma^0 S = \gamma^0, \tag{6.17}$$

that is

$$S^{-1} = \gamma^0 S^\dagger \gamma^0. \tag{6.18}$$

It is easy to show that these matrices form a group. Let S_1 and S_2 be two such matrices. We have

$$(S_1 S_2)^{-1} = S_2^{-1} S_1^{-1} = \gamma^0 S_2^\dagger \gamma^0 \gamma^0 S_1^\dagger \gamma^0 = \gamma^0 S_2^\dagger S_1^\dagger \gamma^0 = \gamma^0 (S_1 S_2)^\dagger \gamma^0. \tag{6.19}$$

Thus, if S_1 and S_2 are Dirac-unitary, then so is their product. The other conditions for group follow trivially from the definition of Dirac-unitary matrices.

These matrices arose from the requirement of keeping the Dirac fields (four spinors) Lorentz invariant. It proved impossible to write the Lorentz transformation of spinors in a unitary form, but it is possible to write them down in a Dirac-unitary form. In that case the Lorentz group structure is left intact. In this article we would like to explore the structure of the most general Dirac-unitary group, and see how it applies to Dirac equation.

Just like in the unitary case, we would like to write Dirac-unitary matrices in an exponential form. Thus, we would like to write $S = \exp(iH)$. In the linearized approximation we have

$$S \approx \mathbf{1} + iH \Rightarrow H = \gamma^0 H^\dagger \gamma^0. \tag{6.20}$$

Matrix H that satisfies condition (4) we would like to call Dirac-Hermitian. An important example of Dirac-Hermitian matrices are the γ^μ matrices. γ^5 on the other hand is not Dirac-Hermitian, but $i\gamma^5$ is. It is easy to see that for a Hermitian matrix H to be Dirac-Hermitian as well we would need to have $[\gamma^0, H] = 0$, i. e. it would need to commute with γ^0.

We can define Dirac conjugation of a matrix A by $\bar{A} \equiv \gamma^0 A^\dagger \gamma^0$. With this definition conditions for Dirac-unitary and Dirac-Hermitian matrices become:

$$U^{-1} = \bar{U}, \tag{6.21}$$
$$H = \bar{H}. \tag{6.22}$$

We can define an inner product of matrices as:

$$(A, B) = \frac{1}{4} tr(\bar{A}B). \tag{6.23}$$

The motivation for this definition is that if we can write A and B as $A = \gamma_\mu A^\mu$, $B = \gamma_\nu B^\nu$, then the inner product of these two matrices is equivalent to the Lorentz vector product of two vectors A^μ and B^ν:

$$\begin{aligned}
(A, B) &= \frac{1}{4} tr(\gamma^0 \gamma_\mu^\dagger A^\mu \gamma^0 \gamma_\nu B^\nu) \\
&= \frac{1}{4} tr(\gamma^0 \gamma_\mu^\dagger \gamma^0 \gamma_\nu) A^\mu B^\nu \\
&= \frac{1}{4} tr(\gamma_\mu \gamma_\nu) A^\mu B^\nu \\
&= \frac{1}{4} 4\eta_{\mu\nu} A^\mu B^\nu \\
&= A^\mu B_\mu.
\end{aligned} \tag{6.24}$$

This inner product also has a desirable feature that it is invariant under

$A \leftarrow U^\dagger A U = A'$, where $U^{-1} = \bar{U}$. We have:

$$
\begin{aligned}
(A', B') &= \frac{1}{4} tr(\gamma^0 U^\dagger A^\dagger U \gamma^0 U^\dagger B U) \\
&= \frac{1}{4} tr(\gamma^0 U^\dagger A^\dagger \gamma^0 B U) \\
&= \frac{1}{4} tr(A^\dagger \gamma^0 B U \gamma^0 U^\dagger) \\
&= \frac{1}{4} tr(A^\dagger \gamma^0 B \gamma^0) \\
&= \frac{1}{4} tr(\gamma^0 A^\dagger \gamma^0 B) \\
&= (A, B).
\end{aligned}
\tag{6.25}
$$

6.4 Representation of Dirac-Unitary Group

In the description of the Standard Model we often deal with various products of Dirac gamma matrices, γ^μ. All the different combinations of products form a Clifford algebra of Dirac matrices. This algebra has a sixteen dimensional basis, and it is conventionally represented with the following set of matrices: $1, \gamma^5, \gamma^\mu, \gamma^5 \gamma^\mu, \sigma^{\mu\nu}$. This choice of basis is based on the transformation of those matrices under the Lorentz group. The above matrices transform as scalar, pseudoscalar, vector, pseudovector, and tensor.

The above representation of Clifford algebra is also related to the basis of the Lie algebra of Dirac-unitary group. We can take the following set of Dirac-Hermitian matrices as our basis: $1/2, i\gamma^5/2, \gamma^\mu/2, \gamma^5\gamma^\mu/2, \sigma^{\mu\nu}/2$. From the Appendix B we can deduce the following commutation relations for this algebra:

$$[\frac{\gamma^\mu}{2}, \frac{\gamma^\nu}{2}] = -i\frac{\sigma^{\mu\nu}}{2},$$

$$[i\frac{\gamma^5}{2}, \frac{\gamma^\mu}{2}] = i\frac{\gamma^5\gamma^\mu}{2},$$

$$[i\frac{\gamma^5}{2}, \frac{\gamma^5\gamma^\mu}{2}] = i\frac{\gamma^\mu}{2},$$

$$[\frac{\gamma^5\gamma^\mu}{2}, \frac{\gamma^\nu}{2}] = \eta^{\mu\nu}\frac{\gamma^5}{2},$$

$$[\frac{\gamma^5\gamma^\mu}{2}, \frac{\gamma^5\gamma^\nu}{2}] = i\frac{\sigma^{\mu\nu}}{2}, \qquad (6.26)$$

$$[i\frac{\gamma^5}{2}, \frac{\sigma^{\mu\nu}}{2}] = 0,$$

$$[\frac{\gamma^\mu}{2}, \frac{\sigma^{\rho\sigma}}{2}] = i(\mathcal{J}^{\rho\sigma})^\mu_\nu\frac{\gamma^\nu}{2},$$

$$[\frac{\gamma^5\gamma^\mu}{2}, \frac{\sigma^{\rho\sigma}}{2}] = i(\mathcal{J}^{\rho\sigma})^\mu_\nu\frac{\gamma^5\gamma^\nu}{2},$$

$$[\frac{\sigma^{\mu\nu}}{2}, \frac{\sigma^{\rho\sigma}}{2}] = i(\mathcal{K}^{\mu\nu\rho\sigma})_{\alpha\beta}\frac{\sigma^{\alpha\beta}}{2},$$

where $\mathcal{J}^{\rho\sigma}_{\mu\nu} = \delta^\rho_\mu\delta^\sigma_\nu - \delta^\rho_\nu\delta^\sigma_\mu$, and $\mathcal{K}^{\mu\nu\rho\sigma}_{\alpha\beta} = \mathcal{J}^{\mu\nu\rho}_\alpha\delta^\sigma_\beta - \mathcal{J}^{\mu\nu\rho}_\beta\delta^\sigma_\alpha$.

6.5   Dirac-unitary gauge theory

In general, Dirac operator $\not{\partial} \equiv \gamma^\mu\partial_\mu$ does not commute with an arbitrary Dirac-unitary transformation. That is

$$[\not{\partial}, U] \neq 0, \qquad (6.27)$$

for an arbitrary U. In order to remedy this, we need to introduce a Dirac-hermitian gauge field A so that we obtain a covariant derivative $\not{\mathcal{D}} = \not{\partial} + i\mathcal{A}$.

Let $U = \exp(-i\theta)$. Where θ is an arbitrary Dirac-hermitian operator. Then in the linearized regime we have the following transformation of the covariant derivative:

$$\not{\partial} + i\mathcal{A} \to \not{\partial} + i\mathcal{A} - i[\not{\partial}, \theta] + [\mathcal{A}, \theta], \qquad (6.28)$$

which implies the following transformation of the gauge field $\mathcal{A}$

$$\mathcal{A} \to \mathcal{A} - [\not{\partial}, \theta] - i[\mathcal{A}, \theta] \qquad (6.29)$$

In order to construct a full dynamical gauge field theory that accomodates the full Dirac-untary group, we would need to build a field strength operator and incorporate it in the full action for this theory. However, it is not immediately clear what would be the best way of constructing this operator. In the case of the non-local gauge field theory we were able to strightforwardly apply the same commutator relations for the covariant derivatives that we used to construct the local gauge theory. In the case of the Dirac-unitary theory we cannot pursue that analogy. We leave this question for the future work.

7 Conclusion and Future Work

7.1 Conclusion

The main two impetuses for my thesis work were the following: incorporating gravity with the other forces arising in the Standard Model, and investigating the mathematical formulations and implications of the non-local unitary gauge theory. These two motivations have led to a unique model of gravity as a component of non-local gauge field theory. This is a natural extension of QED, and the model reduces to QED in the limit where the Planck mass is infinitely large. The evidence that this theory corresctly represents gravity is the following:

1) The motion of a particle in the presence of the gravitational field is exactly described by the geodesic equation (just like in General Relativity), and

2) The field equations for the gravitational field are consistent with the General Relativity in the weak field limit.

The real advantage of this approach to gravity becomes apparent when we try to quantize this theory. The usual obstacles that are present in the attempts to quantize gravity are not present in this theory:

1) The dimensionful interaction constant (Planck mass) is moved from the interaction vertex to gauge field propagator. The new interaction constant is just the dimensionless fine-structure constant. 2) The gravitational field potential is polynomial in fields, so no new adjustments are needed at each stage of the perturbative quantization expansion. 3) In its fully quantized form, the gauge-field propagator has two extra powers of momentum in the denominator, thus making the theory super-renormalizable in 4 dimensions.

7.2   Future Work

We have presented a self-contained theory that on the outset has many desirable qualities: it unifies electromagnetic and gravitational interactions under a single gauge principle, and the corresponding quantum theory is quantiziable. We've demonstrated how the renormalizabuility can be explicitly shown at the one-loop level, however most of the diagrams were much too complicated for a detalied computational analisys. In particular, it would be very desirable to try to find an analogous Ward-Takahashi theorem for this theory and show that it's explicitelly satisfied.

One thing that we have kept emphasizing throughout this work is that within the context of non-local unitary group we can deal with much richer class of propagators than in the ordinary local gauge field theories. This is because in the non-local gauge theory the fundamental field propagators are themselves elements of the Lie algebra of generators of the gauge group. They could, in principle, be as complex as the group itself, which in our case corresponds to a very large group of all possible differential operators. The Dirac field propagator, for instance, has the form

$$S(p, q) = f(\not{p})\delta(p - q), \tag{7.1}$$

which is essentially the same as that of the renormalized propagator. Physically this means that the original propagator is only important insofar as it a) has the correct low-energy form, and b) after the perturbative calculations it leads to the correct physical propagator for all energy scales. It would be interesting if we could find a form of the function $f(\not{p})$ that yields the propagator that is the same after the perturbative renormalization. This would be a "fixed point" of this transformation (not to be confused with the fixed point of Wilson's renormalization group procedure).

Another interesting exercise would be to work with a theory that does not have a Dirac field propagator in the above form. From the point of view of the non-local gaue theory, there is no a-priori reason why the propagator should have the above form, aside from the fact that it's invariant under the Poincaré group transformations. In principle, we could use any Dirac-hermitian differential operator and it would yield a mathematically meaningful theory. Comparing this theory to other Physically accepted theories might lead to some new insight about those, or lead to a new implications for a possible Physics beyond the Standard Model. For instance, in the Standard Model quarks are modelled as essentially the same as the leptons, the only difference being their chromodynamic interaction. Individaul quarks, however, have never been observed and this fact is explained by the theory of quark confinement. That is, the strong force due to its characteristics keeps quarks bound firmly together at an arbitrary length scale. Thus, confinement is explained as a dynamical effect of the strong force.

From the point of view of the non-local gauge field theory we can understand the fact that we observe free leptons by the fact that their propagator commutes with the translation operators. Since the Poincaré group is a subgroup of the non-local gauge group, leptons are free because this subgroup is unbroken in their case. If there were particles out there for which the propagator would not commute with the Poincaré group, then these particles would not be observable as free particles. This might be an interesting way of modelingquarks, in which case confinement could be explained as a purely kinematic effect.

A Group Theory Primer

A.1 Groups

There are many good books that deal with group theory and its applications to Physics [47, 48, 49, 50]. The aim of this appendix is to provide the basic deffinitions of the most essential mathematical concepts that have been used throughout this thesis.

Group. A set G with a multiplication map $* : G \times G \to G$ is called a *group* if the following axioms are satisfied:

G1. *Associativity.* $\forall g_1, g_2, g_3 \in G, g_1 * (g_2 * g_3) = (g_1 * g_2) * g_3$.

G2. *Unit element.* $\exists e \in G$ such that $\forall g \in G, g * e = e * g = g$.

G3. *Inverse elemnt.* $\forall g \in G, \exists h \in G$, such that $g * h = h * g = e$. This h is called the inverse element of g and it is denoted by g^{-1}.

In general, the group multiplication is not commutative. If $g * h = h * g \forall h, g \in G$, then the group is called **abelian**. The **order** of the group is the cardinality of the set G and it is written as $|G|$ or G^o.

Subgroup. Let S be a subset of G. If the group the group multiplication $*$ that is induced by G makes S a group, then S is called a subgroup of G. S is called a *proper subgroup* if it is not equivalent to either G or e.

Coset. Let S be a subgroup of G. We call a subset A of G a *left coset* if $\exists g \in G$, such that $A = \{a | a = g * s, s \in S\}$. A is called a *right coset* if $\exists g \in G$, such that $A = \{a | a = s * g, s \in S\}$.

Theorem 1. Let S be a subgroup of group G. Then each element of G is contained in one and only one left coset of S in G. Furthermore, given any two left cosets of S in G, A and A', there is a one-to-one, onto mapping from the set A to the set A'.

Normal Subgoup. A subgroup N of group G is called a *normal subgroup* if $\forall h \in N, \forall g \in G, ghg^{-1} = h'$, where $h' \in N$.

Thus, N is a normal subgroup if all of its left cosets equal all of its right cosets.

Theorem 2. Let N be a normal subgroup of goup G, and denot by G/N the collection of cosets of N in G. Then for $A, A' \in G/N$ the set $B = A * A' \equiv \{b | b = a * a', a \in A, a' \in A'\}$ is also a coset of N. Thus this product structure makes G/N a group.

A.2 Representations

Representation. A representation of group G is a mapping $R : G \to V$ of the elements of G onto a set of linear operators over a vector space with the following properties:

R1. $R(e) = \mathbf{1}$, where $\mathbf{1}$ is the identity element in the vector space on which the linear operators act.

R2. The group multiplication law is mapped onto the operator multiplication in the vector space on which the linear operators act: $R(g_1)R(g_2) = R(g_1 * g_2) \forall g_1, g_2 \in G$.

The *dimension* of a representation is the dimension of the vector space on which the linear operators $R(g)$ act.

A representation is said to be *reducible* if it has an *invariant subspace*, that is the action of any $R(g)$ on any vector in the subspace is still in the subspace.

A.3 Lie Groups

Lie group. A group G that has a structure of a C^{∞} manifold for which $(g, g') \to g^{-1}g'$ is a C^{∞} map from the manifold $G \times G$ to G is called a *Lie group.*

Lie algebra. A *Lie algebra* is a vector space L over some field F together with bilinear operator $[;] : L \times L \to L$ called the *Lie bracket*, which satisfies the following properties:

L1. Bilinearity: $[ax + by, z] = a[x, z] + b[y, z], [z, ax + by] = a[z, x] + b[z, y]$
   $\forall a, b \in F, \forall, x, y, z \in L.$

L2. $[x, x] = 0, \forall x \in L.$

L3. The *Jacobi identity*: $[x, [y, z]] + [y, [z, x]] + [z, [x, y]] = 0, \forall x, y, z \in L$

Note that *L1* and *L2* imply the anti-symmetry of the Lie bracket:
$[x, y] = -[y, x] \forall x, y \in L.$

Adjoint representation. Let L be a Lie algebra. For $A \in L$, define a linear map $\text{ad}(A)$ on L as

$$\mathrm{A}X = [A, X], \tag{A.1}$$

where $XinL$. This is a Lie algebra homomorphism: for $A, B \in L$

$$[\text{ad}(A), \text{ad}(B)]X = [A, [B, X]] - [B, [A, X]] = [[A, B], X] \tag{A.2}$$

due to Jacobi Identity. Hence, $[\text{ad}(A), \text{ad}(B)] = \text{ad}([A, B])$. Therefore, (ad, L) is a representation of L and is called *adjoint representation*.

Note that

$$\exp[\text{ad}(A)]B = e^A B e^{-A}. \tag{A.3}$$

Let A^i be a basis for a finite, discrete Lie algebra. Then the closure of this algebra implies

$$[A^i, A^j] = if^{ijk}A^k \tag{A.4}$$

for some f^{ijk}. These f^{ijk} are called the structure constants of the algebra. These structure constants uniquely define the adjoint representation. We have

$$(A^i)_{jk} = f_{ijk}. \tag{A.5}$$

A.4   Poincaré Group

Poincaré group. Poincareé group is a Lie group of order ten. Generators of the Poincaré group are labeled P^μ (translations) and $M^{\mu\nu}$ (rigid rotations).

They satisfy the following commutation relations:

$$[P^\mu, P^\nu] = 0, \tag{A.6}$$

$$[M^{\mu\nu}, M^{\rho\sigma}] = i\eta^{\nu\rho}M^{\mu\sigma} - i\eta^{\mu\rho}M^{\nu\sigma} - i\eta^{\nu\sigma}M^{\mu\rho} + i\eta^{\mu\sigma}M^{\nu\rho}, \tag{A.7}$$

$$[M^{\mu\nu}, P^\rho] = -i\eta^{\mu\rho}P^\nu + i\eta^{\nu\rho}P^\mu. \tag{A.8}$$

If the vector space on which the Poncaré group acts is a functional space of fields, then the generators of the group will have the following representation:

$$P^\mu = -i\partial^\mu, \tag{A.9}$$

$$M^{\mu\nu} = i(x^\mu\partial^\nu - x^\nu\partial^\mu) + S^{\mu\nu}, \tag{A.10}$$

where $S^{\mu\nu}$ is the matrix of spin degrees of freedom. For scalar field $S^{\mu\nu} = 0$. For Dirac field we have

$$S^{\mu\nu} = \frac{i}{2}[\gamma^\mu, \gamma^\nu], \tag{A.11}$$

while for Lorentz vectors we have

$$(S^{\mu\nu})_{\alpha\beta} = i(\delta^\mu_\alpha \delta^\nu_\beta - \delta^\mu_\beta \delta^\nu_\alpha). \tag{A.12}$$

A.5 Conformal Group

Conformal Group. Conformal group is an extension of the Poincaré group whose Lie algebra in addition to P^μ and $M^{\mu\nu}$ operators also has operators D (dilatation) and K^μ (special conformal transformation). They satisfy the following commutation relations:

$$[M^{\mu\nu}, K^\lambda] = i(\eta^{\nu\lambda}K^\mu - \eta^{\mu\lambda}), \tag{A.13}$$

$$[K^\mu, P^\nu] = 2i(\eta^{\mu\nu}D - M^{\mu\nu}), \tag{A.14}$$

$$[D, P^\mu] = iP^\mu, \tag{A.15}$$

$$[D, K^\mu] = -iK^\mu, \tag{A.16}$$

$$[K^\mu, K^\nu] = 0, \tag{A.17}$$

$$[M^{\mu\nu}, D] = 0. \tag{A.18}$$

If the vector space on which the Poncaré group acts is a functional space of fields, then the generators of the group will have the following representation:

$$D = -ix^{\mu}\partial_{\mu},$$

(A.19)

$$K^{\mu} = -i(2x_{\mu}x^{\nu}\partial_{\nu} - x^2\partial_{\mu}).$$

(A.20)

B Gamma Matrices

B.1   Pauli Matrices

Pauli matrices, $\sigma_1, \sigma_2, \sigma_3$, are traceless Hermitian two by two matrices. They are generators of the Lie group $SU(2)$. The standard basis for the Pauli matrices is

$$\sigma_1 = \begin{pmatrix} 0 & 1 \\ 1 & 0 \end{pmatrix}, \sigma_2 = \begin{pmatrix} 0 & -i \\ i & 0 \end{pmatrix}, \sigma_3 = \begin{pmatrix} 1 & 0 \\ 0 & -1 \end{pmatrix}. \tag{B.1}$$

B.2   Dirac Gamma Matrices

Dirac gamma matrics, γ^μ are four by four matrices that obey the anticommutation relation

$$\{\gamma^\mu, \gamma^\nu\} = 2\eta^{\mu\nu}. \tag{B.2}$$

In the chiral basis, gamma matrices have the form

$$\gamma^\mu = \begin{pmatrix} 0 & \sigma^\mu \\ \bar{\sigma}^\mu & 0 \end{pmatrix}, \tag{B.3}$$

where

$$\sigma = (1, \sigma), \bar{\sigma} = (1, -\sigma). \tag{B.4}$$

Other important 4×4 matrices are

$$\gamma^5 \equiv i\gamma^0\gamma^1\gamma^2\gamma^3 = -\frac{1}{4!}\epsilon^{\mu\nu\rho\sigma}\gamma_\mu\gamma_\nu\gamma_\rho\gamma_\sigma, \tag{B.5}$$

and

$$\sigma^{\mu\nu} = \frac{i}{2}[\gamma^\mu, \gamma^\nu]. \tag{B.6}$$

These matrices satisfy the following relations:

$$(\gamma^5)^\dagger = \gamma^5,$$

$$(\gamma^5)^2 = \mathbf{1},$$

$$\{\gamma^5, \gamma^\mu\} = \mathbf{0},$$

$$[\gamma^5, \sigma^{\mu\nu}] = \mathbf{0}, \tag{B.7}$$

$$\gamma^5 \sigma^{\mu\nu} = \frac{i}{2} \epsilon^{\mu\nu\sigma\rho} \sigma_{\sigma\rho},$$

$$\not{a}\,\not{b} = a^\mu b_\mu - i\sigma_{\mu\nu} a^\mu b^\nu.$$

In chiral basis we have

$$\gamma^5 = \begin{pmatrix} -1 & 0 \\ 0 & 1 \end{pmatrix} \tag{B.8}$$

B.2.1 Identities involving gamma matrices

Traces of γ matrices can be evaluated as follows:

$$\mathrm{tr}(\mathbf{1}) = 4$$

$$\mathrm{tr}(\text{any odd number of } \gamma\text{s}) = 0$$

$$\mathrm{tr}(\gamma^\mu \gamma^\nu) = 4\eta^{\mu\nu}$$

$$\mathrm{tr}(\gamma^\mu \gamma^\nu \gamma^\rho \gamma^\sigma) = 4(\eta^{\mu\nu}\eta^{\rho\sigma} - \eta^{\mu\rho}\eta^{\nu\sigma} + \eta^{\mu\sigma}\eta^{\nu\rho}) \tag{B.9}$$

$$\mathrm{tr}(\gamma^5) = 0$$

$$\mathrm{tr}(\gamma^\mu \gamma^\nu \gamma^5) = 0$$

$$\mathrm{tr}(\gamma^\mu \gamma^\nu \gamma^\rho \gamma^\sigma \gamma^5) = -4i\epsilon^{\mu\nu\rho\sigma}$$

We are allowed to reverse the order of γ matrices inside the trace:

$$\mathrm{tr}(\gamma^\mu \gamma^\nu \gamma^\rho \gamma^\sigma \cdots) = \mathrm{tr}(\cdots \gamma^\sigma \gamma^\rho \gamma^\nu \gamma^\mu). \tag{B.10}$$

Contractions of γ matrices are given by

$$\gamma^\mu \gamma_\mu = 4$$
$$\gamma^\mu \gamma^\nu \gamma_\mu = -2\gamma^\nu$$
$$\gamma^\mu \gamma^\nu \gamma^\rho \gamma_\mu = 4\eta^{\nu\rho}$$
$$\gamma^\mu \gamma^\nu \gamma^\rho \gamma^\sigma \gamma_\mu = -2\gamma^\sigma \gamma^\rho \gamma^\nu \qquad \text{(B.11)}$$
$$\gamma^\mu \gamma^\nu \gamma^\lambda \gamma^\rho \gamma^\sigma \gamma_\mu = 2(\gamma^\sigma \gamma^\nu \gamma^\lambda \gamma^\rho - \gamma^\rho \gamma^\lambda \gamma^\nu \gamma^\sigma)$$
$$\gamma^\mu \sigma^{\nu\lambda} \gamma_\mu = 0$$
$$\gamma^\mu \sigma^{\nu\lambda} \gamma^\sigma \gamma_\mu = 2\gamma^\sigma \sigma^{\nu\lambda}$$

In d dimensions contractions of gamma matrices are modified and they become:

$$\gamma^\mu \gamma_\mu = d$$
$$\gamma^\mu \gamma^\nu \gamma_\mu = -(d-2)\gamma^\nu$$
$$\gamma^\mu \gamma^\nu \gamma^\rho \gamma_\mu = 4\eta^{\nu\rho} - (4-d)\gamma^\nu \gamma^\rho \qquad \text{(B.12)}$$
$$\gamma^\mu \gamma^\nu \gamma^\rho \gamma^\sigma \gamma_\mu = -2\gamma^\sigma \gamma^\rho \gamma^\nu + (4-d)\gamma^\nu \gamma^\rho \gamma^\sigma$$

We can also contract ϵ tensor:

$$\epsilon^{\alpha\beta\gamma\delta} \epsilon_{\alpha\beta\gamma\delta} = -24$$
$$\epsilon^{\alpha\beta\gamma\mu} \epsilon_{\alpha\beta\gamma\nu} = -6\delta^\mu_\nu \qquad \text{(B.13)}$$
$$\epsilon^{\alpha\beta\mu\nu} \epsilon_{\alpha\beta\rho\sigma} = -2(\delta^\mu_\rho \delta^\nu_\sigma - \delta^\mu_\sigma \delta^\nu_\rho)$$

C Momentum Integrals

Oftentimes when doing momentum integrals it is convenient to combine denominators. We accomplish this by introducing integrals over Feynman parameters. In the case of two denominator factors we have

$$\frac{1}{AB} = \int_0^1 dx \frac{1}{[xA + (1-x)B]^2}.$$
(C.1)

For an arbitrary number of propagator denominators we have

$$\frac{1}{A_1 A_2 \cdots A_n} = \int_0^1 dx_1 \cdots dx_n \delta(\sum x_i - 1) \frac{(n-1)!}{[x_1 A_1 + x_2 A_2 + \cdots + x_n A_n]^n}.$$
(C.2)

An even more general identity is given by

$$\frac{1}{A_1^{m_1} A_2^{m_2} \cdots A_n^{m_n}} = \int_0^1 dx_1 \cdots dx_n \delta(\sum x_i - 1) \frac{\Pi x_i^{m_i - 1}}{[\sum x_i A_i]^{\sum m_i}} \frac{\Gamma(m_1 + \cdots + m_n)}{\Gamma(m_1) \cdots \Gamma(m_n)}.$$
(C.3)

We can make the following table of n-dimensional integrals in Minkowski space:

$$\int \frac{d^n k}{(k^2 + 2k \cdot q - m^2)^\alpha} = \frac{i\pi^{n/2}}{\Gamma(\alpha)(-q^2 - m^2)^{\alpha - n/2}} \Gamma(\alpha - n/2)$$

$$\int \frac{d^n k\, k_\mu}{(k^2 + 2k \cdot q - m^2)^\alpha} = \frac{-i\pi^{n/2}}{\Gamma(\alpha)(-q^2 - m^2)^{\alpha - n/2}} q_\mu \Gamma(\alpha - n/2)$$

$$\int \frac{d^n k\, k_\mu k_\nu}{(k^2 + 2k \cdot q - m^2)^\alpha} = \frac{i\pi^{n/2}}{\Gamma(\alpha)(-q^2 - m^2)^{\alpha - n/2}} \Big[q_\mu q_\nu \Gamma(\alpha - n/2) + \frac{1}{2}\eta_{\mu\nu}(-q^2 - m^2)\Gamma(\alpha - 1 - n/2) \Big]$$

$$\int \frac{d^n k\, k_\mu k_\nu k_\lambda}{(k^2 + 2k \cdot q - m^2)^\alpha} = \frac{i\pi^{n/2}}{\Gamma(\alpha)(-q^2 - m^2)^{\alpha - n/2}} \Big[- q_\mu q_\nu q_\lambda \Gamma(\alpha - n/2) - \frac{1}{2}(\eta_{\mu\nu}q_\lambda + \eta_{\nu\lambda}q_\mu + \eta_{\lambda\mu}q_\nu) \times (-q^2 - m^2)\Gamma(\alpha - 1 - n/2) \Big]$$

$$\int \frac{d^n k\, k_\mu k_\nu k_\lambda k_\rho}{(k^2 + 2k \cdot q - m^2)^\alpha} = \frac{i\pi^{n/2}}{\Gamma(\alpha)(-q^2 - m^2)^{\alpha - n/2}} \Big[q_\mu q_\nu q_\lambda q_\rho \Gamma(\alpha - n/2) + \frac{1}{2}(q_\mu q_\nu \eta_{\lambda\rho} + \text{perm})(-q^2 - m^2)\Gamma(\alpha - 1 - n/2) + \frac{1}{4}(\eta_{\mu\nu}\eta_{\lambda\rho} + \text{perm})(-q^2 - m^2)\Gamma(\alpha - 2 - n/2) \Big]$$

(C.4)

A general formula has the following form:

$$\int \frac{d^n k\, k_{\mu_1} k_{\mu_2} \cdots k_{\mu_p}}{(k^2 + 2k \cdot q - m^2)^\alpha} = \frac{i\pi^{n/2}}{\Gamma(\alpha)(-q^2 - m^2)^{\alpha - n/2}} T_{\mu_1 \mu_2 \cdots \mu_p}, \qquad \text{(C.5)}$$

where:

$$T_{\mu_1\mu_2\cdots\mu_p} = (-1)^p \Big[q_{\mu_1}q_{\mu_2}\cdots q_{\mu_p}\Gamma(\alpha - n/2)$$

$$+ \frac{1}{2}\sum_{\text{perm}}(\eta_{\mu_1\mu_2}q_{\mu_3}\cdots q_{\mu_p})(-q^2 - m^2)\Gamma(\alpha - 1 - n/2)$$

$$+ \frac{1}{4}\sum_{\text{perm}}(\eta_{\mu_1\mu_2}\eta_{\mu_3\mu_4}q_{\mu_5}\cdots q_{\mu_p})(-q^2 - m^2)\Gamma(\alpha - 2 - n/2) \tag{C.6}$$

$$\vdots$$

$$+ 2^{-p/2}\sum_{\text{perm}}(\eta_{\mu_1\mu_2}\eta_{\mu_3\mu_4}\cdots\eta_{\mu_{p-1}\mu_p})$$

$$\times (-q^2 - m^2)^{p/2}\Gamma(\alpha - p/2 - n/2)\Big]$$

for p even. For p odd, the last term should be:

$$+ 2^{-[p/2]}\sum_{\text{perm}}(\eta_{\mu_1\mu_2}\cdots\eta_{\mu_{p-2}\mu_{p-1}}q_{\mu_p})$$

$$\times (-q^2 - m^2)^{[p/2]}\Gamma(\alpha - [p/2] - n/2), \tag{C.7}$$

where $[p]$ means taking the largest integer not greater than p.

By contracting various k_μ we can also derive a succession of related formulas involving k^2.

We can obtain corresponding formulas in Euclidian space by substituting $-m^2$ for m^2 and 1 for i.

D Feynman Rules (Tree level)

D.1 Non-local ϕ^4 theory

D.1.1 Propagator

$p \longrightarrow q$ $\qquad S(p,q) = \frac{i}{p^2-m^2+i\epsilon}\delta(p-q)$

D.1.2 Vertex

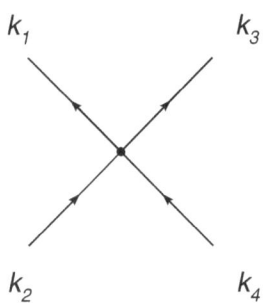

$$D(k_1,k_2,k_3,k_4) = \frac{-i\lambda}{[1+L_1^2(\sum_i k_i^2)+L_2^2(\sum_{i\neq j} k_i\cdot k_j)]^\alpha}\delta(k_1-k_2+k_3-k_4)$$

D.2 Non-Local Gauge Field Theory

D.2.1   Fermion Propagator

p ———————→——— q $S(p,q) = \frac{i(\not{p}+m)}{p^2-m^2+i\epsilon}\delta(p-q)$

D.2.2   Ghost Propagator

p ·············→········· q $C(p,q) = \frac{1}{p^2+i\epsilon}\delta(p-q)$

D.2.3   Gauge Field Propagator

$\mu,\, k_1,\, k_2$ 〜〜〜〜〜〜 $\nu,\, k_3,\, k_4$

$$
\begin{aligned}
D^{\mu\nu}(k_1,k_2,k_3,k_4) = \; & \frac{-i}{(k_1-k_2)^2+i\epsilon} \\
& \times \left(\eta_{\mu\nu} - (1-\zeta)\frac{(k_1-k_2)^\mu(k_1-k_2)^\nu}{(k_1-k_2)^2} \right) \\
& \times \frac{1}{1+\frac{k_1\cdot k_3}{\alpha M_{Pl}^2}}\delta(k_1-k_2+k_3-k_4),
\end{aligned}
$$

D.2.4   Fermion - Gauge Boson Vertex

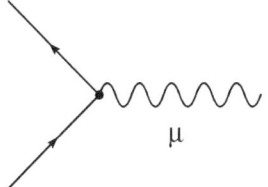

μ

$ie\gamma^\mu$

D.2.5 Ghost - Gauge Boson Vertex

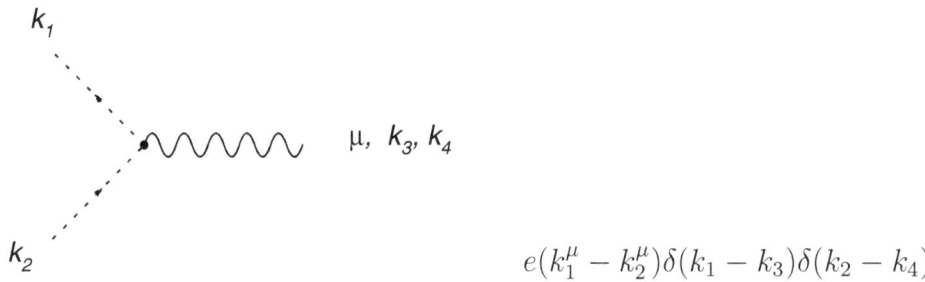

$$e(k_1^\mu - k_2^\mu)\delta(k_1 - k_3)\delta(k_2 - k_4)$$

D.2.6 3-Boson Vertex

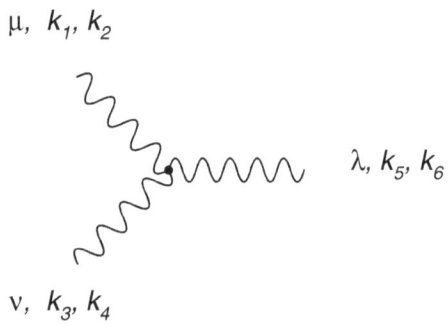

$$
\begin{aligned}
i\Gamma^{\mu\nu\lambda}(k_1, k_2, k_3, k_4, k_5, k_6) = ie(& K(k_1, k_2, k_3, k_4, p, q, k_5, k_6)p^\lambda \delta(p - q)\eta^{\mu\nu} \\
+ & K(k_3, k_4, k_5, k_6, p, q, k_1, k_2)p^\mu \delta(p - q)\eta^{\nu\lambda} \\
+ & K(k_5, k_6, k_1, k_2, p, q, k_3, k_4)p^\nu \delta(p - q)\eta^{\mu\lambda})
\end{aligned}
$$

D.2.7 4-Boson Vertex

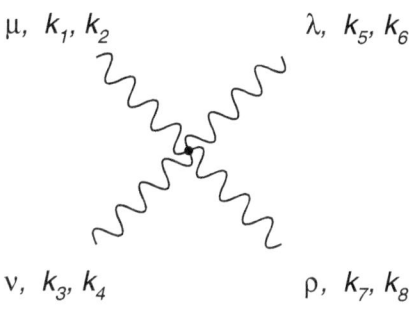

μ, k_1, k_2 λ, k_5, k_6

ν, k_3, k_4 ρ, k_7, k_8

$$i\Gamma^{\mu\nu\lambda\rho}(k_1, k_2, k_3, k_4, k_5, k_6, k_7, k_8)$$

$$= ie^2[K(k_1, k_2, k_3, k_4, k_5, k_6, k_7, k_8)(\eta^{\mu\lambda}\eta^{\nu\rho} - \eta^{\mu\rho}\eta^{\nu\lambda})$$

$$+ K(k_1, k_2, k_5, k_6, k_3, k_4, k_7, k_8)(\eta^{\mu\nu}\eta^{\lambda\rho} - \eta^{\mu\rho}\eta^{\lambda\nu})$$

$$+ K(k_1, k_2, k_7, k_8, k_5, k_6, k_3, k_4)(\eta^{\mu\lambda}\eta^{\rho\nu} - \eta^{\mu\nu}\eta^{\rho\lambda})]$$

In the above expressions we have

$$K(k_1, k_2, k_3, k_4, k_5, k_6, k_7, k_8) = [\delta(k_8 - k_1)\delta(k_4 - k_5)\delta(k_2 - k_3)\delta(k_6 - k_7)$$

$$- \delta(k_8 - k_3)\delta(k_2 - k_5)\delta(k_4 - k_1)\delta(k_6 - k_7)$$

$$- \delta(k_7 - k_4)\delta(k_1 - k_6)\delta(k_2 - k_3)\delta(k_8 - k_5)$$

$$+ \delta(k_6 - k_3)\delta(k_2 - k_7)\delta(k_4 - k_1)\delta(k_8 - k_5)]$$

References

[1] H. S. Snyder, Phys Rev. **71**, 38 (1947).

[2] H. Yukawa, Phys. Rev. **76**, 300 (1949).

[3] H. Yukawa, Phys. Rev. **80**, 1047 (1950).

[4] C. N. Yang and R. L. Mills, Phys. Rev. **96**, 191 (1954).

[5] R. Utiyama, Phys. Rev. **101**, 1597 (1956).

[6] J. A. Wheeler, Ann. Phys., **2**, 604 (1957).

[7] T. W. B. Kibble, J. Math. Phys., **2**, 212 (1961).

[8] D. A. Kirzhnits, Sov. Phys. JETP, **14**, 395 (1962).

[9] D. A. Kirzhnits, Sov. Phys. JETP, **18**, 103 (1964).

[10] D. A. Kirzhnits, Sov. Phys. JETP, **18**, 1390 (1964).

[11] H. Steudel, Z. Naturf., **21a**, 1826 (1966).

[12] A. L. Birch and J. S. Dowker, J. Phys. A, **2**, 624 (1969).

[13] T. Takabayasi, Prog. Theor. Phys., **48**, 1718 (1972).

[14] S. W. Hawking, Nucl. Phys. **B 144**, 349 (1978)

[15] A. Connes, J. Lett. Nucl. Phys. B (Proc Suppl.) **18**, 29 (1990).

[16] A. Connes, *Non Commutative Geometry*, Academic Press, London (1994).

[17] A. Connes, J. Math. Phys. **36** 6194 (1995).

[18] A. Connes, hep-th/9603053.

[19] R. Brout, hep-th/9706200.

[20] P. Gibbs, hep-th/9506171.

[21] Z. Dongpei, Physics Scripta **34**, 738 (1986).

[22] J. W. Moffat, Phys. Rev. **D 41**, 1177 (1990).

[23] P. A. M. Dirac, *General Theory of Relativity*, (Princeton University Press, Princeton, 1996)

[24] C. W. Misner, K. S. Thorne, J. A. Wheeler, *Gravitation*, (W. H. Freeman and Company, New York, 1973)

[25] R. M. Wald, *General Relativity*, (University of Chicago press, Chicago, 1984)

[26] B. F. Schutz, *A First Course in General Relativity*, (Cambridge University Press, Cambridge, 1985)

[27] B. F. Schutz, *Geometrical Methods of Mathematical Physics*, (Cambridge University Press, Cambridge, 1980)

[28] S. Weinberg, *Gravitation and Cosmology*, (John Wiley & Sons, 1972)

[29] A. Einstein, "Die Grundlage der allgemeinen Relativitätstheorie," Annalen der Physik, **49**, 1916.

[30] M. J. Duff, Kaluza-Klein Theory in Perspective, hep-th/9410046.

[31] N. D. Birrel and P. C. W. Davis, *Quantum Fields in Curved Space* (Cambridge University Press, Cambridge, 1982).

[32] J. Z. Simon, Phys. Rev. **D 41**, 3720 (1990).

[33] R. M. Wald, *Quantum Field Theory in Curved Spacetime and Back Hole Thermodynamics*, (The University of Chicago Press, Chicago and London, 1994)

[34] C. M. Will, gr-qc/0103036

[35] C. M. Will, gr-qc/9811036

[36] S. Carlip, gr-qc/0108040

[37] C. Rovelli, gr-qc/9903045

[38] C. Rovelli, gr-qc/0006061

[39] J. Butterfield and C. J. Isham, gr-qc/9901024

[40] J. Butterfield and C. J. Isham, gr-qc/9903072

[41] V. O. Rivelles, hep-th/0304073

[42] E. Fermi, Il Nuovo Cimento, **11**, 1 (1934).

[43] E. Fermi, Zeitschrift für Physik, **88**, 161 (1934).

[44] S. L. Glashow, Nucl. Phys. **22**, 579 (1961).

[45] S. Weinberg, Phys. Rev. Lett. **19**, 1264 (1967).

[46] A. Salam, Elementary Particle Theory, ed N. Svartholm, Stockholm: Almqvist and Wiksell (1968).

[47] Y. Oono, Group Theory - elements behind applications, unpublished (1994).

[48] H. Georgi, Lie Algebras in Perticle Physics, Perseus Books (1999).

[49] S. Lang, Algebra, Addison-Wesley (1997).

[50] R. Geroch, Mathematical Physics, The University of Chicago Press (1985).

[51] R. P. Feynam, Feynman Lectures on Gravitation, Addison Wesley Longman (1995).

[52] R. P. Feynman, Acta Physics Polonica **24**, 697 (1963).

[53] R. M. Wald, Phys. Rev. D **33**, 3613 (1986).

[54] C. Brans and R.H Dicke, Phys. Rev. **124**, 925 (1961).

[55] P. G. Bergmann, Int. J. Theor. Phys. **1**, 25 (1968).

[56] A. D. Sakharov, Sov. Phys. Dokl. **12**, 1040 (1968).

[57] C. Barcelo, S. Liberati, M. Visser, Living Rev. Rel. **8**, 12 (2005)

[58] J. D. Bekenstein, astro-ph/0403694 (2004).

[59] M. Milgrom, Astrophysical Journal (2002).

[60] K. G. Wilson, Rev. Mod. Phys. **47**, 4, 773 (1975).

[61] D. Bailin & A. Love, *Introduction to Gauge Field Theory*, IOP Publishing, London, 1993.

[62] M. Kaku, *Quantum Field Theory*, Oxford University Press, New York 1993.

[63] M. E. Peskin & D. V. Schroeder, *An Introduction to Quantum Field Theory*, Addison-Wesley Publishing, Reading, 1995.

[64] D. Griffiths, *Introduction to Elementary Particles*, John Wiley & Sons, 1987.

[65] P. Ramond, *Field Theory A Modern Primer*, The Benjamin/Cummings Publishing Company, Reading, 1981.

Vita

Bojan Tunguz was born in Sarajevo, Bosnia and Herzegovina, in 1974. In 1992 he and his family moved to Croatia, and he is now the citizen of that country. He attended Stanford University where he obtained a B.S. in Physics in 1997 and an M.S. in Applied Physics in 1999. Bojan enjoys reading, writing, watching movies and weightlifting.